Convergence of
Blockchain, AI, and IoT

Innovations in Big Data and Machine Learning

Series Editors:
Rashmi Agrawal and Neha Gupta

This series will include reference books and handbooks that will provide the conceptual and advanced reference materials that cover building and promoting the field of Big Data and machine learning, which will include theoretical foundations, algorithms and models, evaluations and experiments, applications and systems, case studies, and applied analytics in specific domains, or on specific issues.

Artificial Intelligence and Internet of Things
Applications in Smart Healthcare
Edited by Lalit Mohan Goyal, Tanzila Saba, Amjad Rehman, and Souad Larabi

Reinventing Manufacturing and Business Processes through Artificial Intelligence
Edited by Geeta Rana, Alex Khang, Ravindra Sharma, Alok Kumar Goel, and Ashok Kumar Dubey

Convergence of Blockchain, AI, and IoT
Concepts and Challenges
Edited by R. Indrakumari, R. Lakshmana Kumar, B. Balamurugan, and Vijanth Sagayan Asirvadam

Exploratory Data Analytics for Healthcare
Edited by R. Lakshmana Kumar, R. Indrakumari, B. Balamurugan, and Achyut Shankar

For more information on this series, please visit https://www.routledge.com/Innovations-in-Big-Data-and-Machine-Learning/book-series/CRCIBDML

Convergence of Blockchain, AI, and IoT
Concepts and Challenges

Edited by
R. Indrakumari, R. Lakshmana Kumar,
B. Balamurugan, and Vijanth Sagayan Asirvadam

CRC Press
Taylor & Francis Group
Boca Raton London New York

CRC Press is an imprint of the
Taylor & Francis Group, an **informa** business

First edition published 2022
by CRC Press
6000 Broken Sound Parkway NW, Suite 300, Boca Raton, FL 33487-2742

and by CRC Press
4 Park Square, Milton Park, Abingdon, Oxon, OX14 4RN

© 2022 selection and editorial matter, R. Indrakumari, R. Lakshmana Kumar, B. Balamurugan, and Vijanth Sagayan Asirvadam; individual chapters, the contributors

CRC Press is an imprint of Taylor & Francis Group, LLC

Library of Congress Cataloging-in-Publication Data
A catalog record for this title has been requested

ISBN: 978-0-367-53264-2 (hbk)
ISBN: 978-0-367-53265-9 (pbk)
ISBN: 978-1-003-08118-0 (ebk)

DOI: 10.1201/9781003081180

Typeset in Times
by codeMantra

Contents

Preface

The success of blockchain, artificial intelligence, and the Internet of things is purely depending on the security provided to data. The convergence of blockchain, Big Data, and IoT presented in this book depicts the current and future innovations in these fields. The future enhancements with the convergence of these three technologies will change economic growth with digital currency and clinical diagnosis in a wider sense. This book also enhances strategic technological trends through spectrum computing and smart contract techniques.

In Chapter 1, Indrakumari, Vijayalakshmi, Shraddha Sagar, and Poongodi demystify the affiliation of blockchain, artificial intelligence, and the Internet of things in a broad manner.

In Chapter 2, L. Godlin Atlas, Arjun K.P., and Bindu Babu introduced a new blockchain architecture that is recommended for hospitals and clinics, which provides control to patients to increase their accessibility of medical records. They also show how to enhance the security of records with the power of blockchain and cloud-based structures.

In Chapter 3, Dr. G S Pradeep Ghantasala, Anuradha Reddy, and M Arvindhan narrate the innumerable Big Data, IoT, and blockchain-distributed hyperledger framework.

In Chapter 4, Saravanan et al. investigate the peer-to-peer cryptographic secured blockchain technology for IoT and Big Data in healthcare applications such as collecting the data, analyzing the data, and providing full-pledged care to the patients by the healthcare management systems by the integration of IoT and Big Data to provide high privacy, authentication, integrity, and non-repudiation to the applications.

In Chapter 5, Nirbhay et al. give a basic introduction to "Big Data" and its application to draw attention to process data while maintaining the reliability of the data. Finally they have conclude this chapter , in which the main focus was to visualize and create a relationship between process data, reliability data, and Big Data for the era of Industry (I.4).

In Chapter 6, Gayathri et al. discuss the possibilities of tracking the IoT devices and other objects where a blockchain environment shall be adopted to take the entire world into a transparent and quality lifestyle for everyone within 2030.

In Chapter 7, Revathy et al. show the higher stability of energy comparison using modified Jaya Optimization algorithm when compared to the other techniques. The Jaya optimization algorithm can be slightly modified and updated for QoS parameters such as throughput, error rate, jitter, and scalability.

In Chapter 8, Shiva Sankar et al. talk about the convergence of blockchain, IoT, and AI and the advantages regarding security by digital identification numbers and cryptography.

In Chapter 9, Peter Soosai Anandaraj et al. discuss various essential criteria for the confluence of blockchain and AI technology, which will aid in the formation of a sustainable and eco-friendly smart society.

In Chapter 10, Kavita Kumari et al. give a basic introduction to blockchain and its functionalities in healthcare and an overview of counterfeit drugs.

MATLAB® is a registered trademark of The MathWorks, Inc. For product information, please contact:
The MathWorks, Inc.
3 Apple Hill Drive
Natick, MA 01760-2098 USA
Tel: 508-647-7000
Fax: 508-647-7001
E-mail: info@mathworks.com
Web: www.mathworks.com

Editors

Indrakumari Ranganathan is working as an assistant professor in the School of Computing Science and Engineering, Galgotias University, NCR Delhi, India. She has completed an M.Tech in Computer and Information Technology from Manonmaniam Sundaranar University, Tirunelveli. Her main thrust areas are Big Data, Internet of Things, data mining, data warehousing, and its visualization tools such as Tableau and QlikView. She has several top-notch conferences in his resume and has published over 45 quality journal, conference, and book chapters combined.

Lakshmana Kumar Ramasamy is working as head of the Center of Excellence for Artificial Intelligence and Machine Learning at Hindusthan College of Engineering and Technology, Tamil Nadu, India. He is currently pursuing his postdoctoral fellowship at Thu Dau Mot University, Vietnam. He is also working as an offshore director for Research & Development (AI) in a Canadian-based company (ASIQC) in the Vancouver region of British Columbia, Canada. He represents the Technical Group Committee for the National Cyber Defence Research Centre (NCDRC), Government of India. He is the founding member of IEEE SIG on Big Data for Cyber Security and Privacy, IEEE. He serves as a core member in the Editorial Advisory Board of Artificial Intelligence group in Cambridge Scholars Publishing, UK. He is a member of IEEE. He was invited as a keynote speaker for Asia Artificial Intelligence Virtual Summit 2020 (AVIS 2020), which is Asia's first biggest virtual summit on artificial intelligence held in Malaysia in June 2020. He is a global chapter lead for Machine Learning for Cybersecurity (MLCS). He is involved in research, and has expertise, in AI and blockchain technologies.

Balamurugan Balusamy has served up to the position of associate professor in his stint of 14 years of experience with VIT University, Vellore. He has completed his bachelor's, master's, and PhD from premier institutions. His passion is teaching, and he adapts different design thinking principles while delivering his lectures. He has published around 30 books on various technologies and visited 15+ countries for his technical discourse. He has several top-notch conferences in his resume and has published over 150 quality journal, conference, and book chapters combined. He serves in the advisory committee for several start-ups and forums and does consultancy work for the industry on Industrial IoT. He has given over 175 talks at various events and symposiums. He is currently working as a professor at Galgotias University and teaches students, and he does research on blockchain and IoT.

Vijanth Sagayan Asirvadam is from an old mining city in Malaysia called Ipoh. He studied at the University of Putra, Malaysia, for Bachelor of Science (Hon) majoring in Statistics and graduated in April 1997 before leaving for Queen's University Belfast to do his master's where he received his Master of Science degree in Engineering Computation with distinction. He has worked briefly as a lecturer in a private higher institution and as a system engineer in Multimedia University, Malaysia. He later joined the Intelligent Systems and Control research group at Queen's University Belfast in November 1999 where he completed his doctorate (PhD) research in Online and Constructive Neural Learning Methods. He is currently a lecturer in the Faculty of Information Science and Technology at Multimedia University, Malaysia. His research interests include neural network and statistical learning for black-box modeling, model validation, and data mining.

Contributors

A. Ambikapathy
Department of EEE
Galgotias College of Engineering
Greater Noida, India

K. P. Arjun
School of Computing Science
    and Engineering
Galgotias University
Greater Noida, India

M. Arvindhan
School of Computing Science
    and Engineering
Galgotias University
Greater Noida, India

Vijanth S. Asirvadam
Department of Electrical and Electronic
    Engineering
Universiti Teknologi PETRONAS
Seri Iskandar, Malaysia

Bindu Babu
ECE
Easwari Engineering College,
    Chennai, India

B. Balamurugan
School of Computer Science
    and Engineering
Galgotias University
Greater Noida, India

Nopasit Chakpitak
Department of Computer Science
International College of Digital
    Innovation - Chiang Mai University
Chiang Mai, Thailand

Ahmad Faraz
Department of EEE
Galgotias College of Engineering
    and Technology
Greater Noida, India

S. P. Gayathri
Department of Computer Science
    and Applications
The Gandhigram Rural Institute
    (Deemed to be University)
Gandhigram, India

L. Godlin Atlas
School of Computing Science
    and Engineering
Galgotias University
Greater Noida, India

A. Ilavendhan
Veltech Rangarajan Dr. Sagunthala
    R&D Institute of Science
    & Technology
Chennai, India

R. Indrakumari
School of Computing
    Science and Engineering
Galgotias University
Greater Noida, India

P. Kavitha Rani
Sri Krishna College of Engineering
    and Technology
Coimbatore, India

Deepak Kumar Saini
ST Microelectronics
Greater Noida, India

Kavita Kumari
School of Computing Science
    & Engineering
Galgotias University
Greater Noida, India

Nirbhay Mathur
Department of Electrical
    and Electronic
    Engineering
Universiti Teknologi
    PETRONAS
Seri Iskandar,
    Malaysia

A. Peter Soosai Anandaraj
Veltech Rangarajan
    Dr. Sagunthala R&D Institute
    of Science & Technology
Chennai, India

T. Poongodi
School of Computing Science
    and Engineering
Galgotias University
Greater Noida, India

G S Pradeep Ghantasala
Department of Computer Science
    and Engineering
Chitkara University
Rajpura, India

Anuradha Reddy
Department of Computer Science
    and Engineering
Malla Reddy Institute of Technology
    and Science
Hyderabad, India

G. Revathy
School of Computing
SASTRA University
Thanjavur, India

Shraddha Sagar
School of Computing Science
    and Engineering
Galgotias University
Greater Noida, India

Kavita Saini
School of Computing Science
    & Engineering
Galgotias University
Greater Noida, India

T. Saravanan
Department of CSE
St.Martin's Engineering College
Secundearabad, India

K. Selvakumar
Department of Information Technology
Annamalai University
Chidambaram, India

R. Sendhil
Department of Computer Science
    & Technology
Madanapalle Institute of Technology
    & Science
Madanapalle, India

Himanshu Singh
Department of EEE
Galgotias College of Engineering
    and Technology
Greater Noida, India

Siva Shankar Ramasamy
Department of Computer Science
International College of Digital
    Innovation - Chiang Mai University
Chiang Mai, Thailand

S. Vijayalakshmi
Department of Data Science
CHRIST (Deemed to be University)
Pune, India

1 Affiliation of Blockchain, IoT, and Big Data
Demystified

R. Indrakumari
Galgotias University

S. Vijayalakshmi
CHRIST (Deemed to be University)

Shraddha Sagar and T. Poongodi
Galgotias University

CONTENTS

DOI: 10.1201/9781003081180-1

1

1.1 INTRODUCTION

The digital revolution shifted the global manufacturing processes with significant advancements in technologies that leverage production and its associated decisions. Blockchain, the Internet of Things (IoT), and Artificial Intelligence (AI) are ubiquitous technologies that promote the capabilities in an operational process that needs renovation. This digital transformation accelerates a huge impact on the convergence of these technologies in a strategic way by keeping in mind of the present and the future. IoT enables sensors, actuators, physical objects, virtual objects, platforms, networks, people, and services to be interconnected for sharing their data independently. The "things" in IoT gather and transfer data from the surrounding environment. IoT is a driving force for Industry 4.0 because of triggering technological update that spans a broad range of domains.

According to Gartner, it is predicted that 20.8 billion connected devices will be exploited worldwide by 2020. This would surely bring exciting opportunities in the human life that greatly improves safety, efficiency, and productivity in numerous business sector. Blockchain holistically manages the relationship among participants while sharing the data source in financial transactions. Accountability and security are also major concerns in addition to compliance with regulations, resulting in time delay reduction, better consistency, minimized risks, and improved quality.

IoT industrializes various real-world applications, including smart cities, smart transportation to promote human life more reliable. AI supports companies to get closer to their customers, enriches employee's productivity, and accelerates innovation. IoT connects all smart objects or devices, blockchain guarantees end-to-end security, and AI takes IoT systems to behave intelligently. The integration of blockchain, IoT, and AI makes digital transformation more flexible and easier. A massive amount of data is sensed and generated from several sensor devices in the IoT environment. AI plays a significant role in analyzing the Big Data generated by IoT devices and delivers an accurate analysis of real-time data. Blockchain implements a decentralized architecture, and it affords secure sharing of resources among participating nodes in the network. It overcomes the challenges faced with AI techniques by not incorporating a centralized blockchain, AI, and IoT, which are recognized as a novel technology that can enhance the business process and disrupt the entire industry. Blockchain provides a decentralized and shared distributed ledger to enhance security, trust privacy, and transparency. A blockchain is similar to a register or ledger that stores all kinds of assets [1].

AI detects the pattern and outcomes of the business process [2]. IoT drives the automatization of industries and the user-friendliness of business processes. These three technologies can be converged to get innovative outcomes. IoT gathers data, blockchain gives the infrastructure, while AI optimizes processes and rules [3]. These three technologies can exploit their full potential if converged.

1.2 BLOCKCHAIN

Blockchain is a revolutionizing technology that changes the internetworking system. The main features of blockchain technology are [4]:

1. Blockchain creates a series of records to store data where the new incoming records are accommodated in a block with a link to the previous record. This technique of linking the record gave the name blockchain.
2. Distributed ledger concept is used here, which makes the record more transparent.
3. Blockchain follows the cryptography concept to secure user-valuable information, and the distributed ledger prevents hacking the data and hence is considered as the backbone of cryptocurrency.
4. Peer-to-peer sharing is allowed without a mediator.
5. Irrespective of the content what the user shares, the blockchain network will retain ownership of the content unless the owner sells it to someone.
6. Private Key cryptography is used to protect personal information.

Blockchain technology allows user to share their content securely without the intervention of the mediator or a central governing system.

1.2.1 BLOCKS

Blocks are considered as the heart of the blockchain technology, and a data file is used to record the transaction on the network [5]. The blocks are connected to the chain securely, which makes it impossible to tamper with it. For instance, if someone bought the bitcoins, they will be provided with a private key, which is the digital signature. If the bitcoin is to be purchased, the persons should provide the private key that is verified by the bitcoin miner. If the person used the bitcoin, the transaction details are saved in a new block and linked to the previous transaction with a series of characters (Figure 1.1).

1.2.2 HASHES

One of the reasons the blockchain is so popular is that they contain information; although distributed, it is highly encrypted. Data on the blockchain are under encryption by creating a hash. An algorithm is required to create a hash, and it acts by taking the transaction information and converting it to a series of numbers and letters [6]. Hashes are always of the same length.

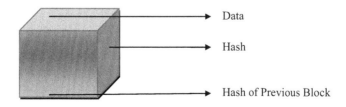

FIGURE 1.1 Blocks.

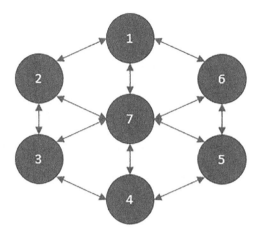

FIGURE 1.2   Nodes.

1.2.3   NODES

Nodes are the basic infrastructure of the blockchain, which can be any kind of device like a server, laptop, or computer. Nodes are focal points of activity spread all over the blockchain network. The copies of the blocks, transaction details, and records are available in the nodes. All the nodes in the chain are connected to exchange the latest data to each other as shown in figure 1.2; hence, the nodes are always up to date. A device that contains the full copy of the transaction history is called the full node.

1.2.4   TYPES OF BLOCKCHAIN

Blockchain is an encrypted repository of digital information. A blockchain has a decentralized and distributed style of a network of computers. The three types of *blockchain* are as follows [7] (Table 1.1):

- *Public Blockchain*
- *Private Blockchain*
- *Consortium or Federated Blockchain*

1.2.5   BLOCKCHAIN PRODUCTS

1.2.5.1   Bitcoin

Bitcoin is the base and stimulus for all blockchain systems, first described by Satoshi Nakamoto in 2008 [8] and announced as open-source software in the year 2009. Bitcoin uses the blockchain concept to evade the double-spending concept in the transaction. It is a peer-to-peer technology that is not governed by any central authority or banks. This is a virtual currency that is highly secured. Every transaction is stored by the bitcoin protocol using a general ledger. The bitcoins are stored as digital credentials in the wallet that use the public-key encryption (Figure 1.3).

TABLE 1.1

Public VS Private VS Consortium *Blockchain*

Public Blockchain	Private Blockchain	Consortium or Federated Blockchain
Everyone can run BTC/ LTC full node	Not possible for all members to run a full node	Selected members of the consortium can run a full node
Everyone can make transactions	Not possible for all members to make transactions	Selected members of the consortium can make transactions
Everyone can review/ audit the blockchain	Not possible for all members to review/audit the blockchain	Selected members of the consortium can review/audit the blockchain
Example: Bitcoin, Litecoin, etc.	Example: Bankchain	Example: r3, EWF

1.3 ARTIFICIAL INTELLIGENCE

Over the most recent quite a long while, several researchers have shown an interest in AI due to its increasing use. Funding interests in organizations creating and commercializing AI items and innovation have surpassed $2 billion since 2011 [9]. Innovation organizations have contributed billions all the more acquiring AI start-ups. Press inclusion of the subject has been winded, filled by the gigantic speculations and by intellectuals attesting that PCs are beginning to execute occupations, will be smarter than individuals long before, and could hamper the survival of mankind.

1.3.1 INTRODUCTION TO ARTIFICIAL INTELLIGENCE

Since the innovation of computers or machines, their ability to perform different tasks continued developing very fast. People have built up the intensity of

FIGURE 1.3 Bitcoin.

computer systems regarding their different working areas, high-intensity speed, and reducing size according to the time. AI is one the domains of computer science, which creates the computer system and machines that are more intelligent than the humans.

John McCarthy, who is the father of Artificial Intelligence, has defined AI as "The science and engineering of making intelligent machines, especially intelligent computer programs". AI is the new buzzword of the 21st century that makes computers, computer-controlled robots, or the intelligently thinking software packages comparable to the thinking of intelligent humans.

AI has been achieved by studying that how the human mind thinks, learns, decides, and works for solving a problem. After studying the outcomes of this study, an intelligent software and system can be developed. The main aim of AI is to create expert systems and to implement human intelligence in machines for making the system understand, think, learn, and behave like a human being (Figure 1.4).

AI is the science and technology that consist of different disciplines such as Computer Science, Biology, Psychology, Linguistics, Mathematics, and Engineering. AI aims to do the development of the functions for computers that are related to human intelligence like reasoning, learning, and problem-solving.

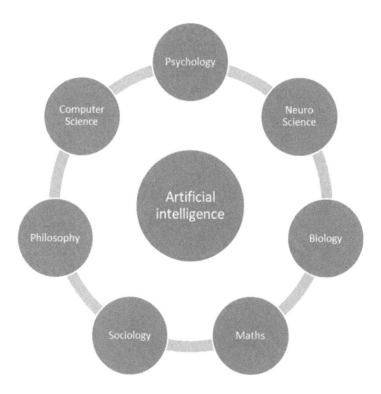

FIGURE 1.4 Domains of Artificial Intelligence.

1.3.2 CATALYSTS OF PROGRESS

By the end of 2000s, various factors have led to the progress in AI, especially in few important techniques. Various factors that lead to the recent progress are Moore's law, Big Data, the internet, and the cloud, and new algorithms.

The persistent expansion in computing power accessible at a given cost and size, in some cases known as Moore's law after the fonder of Intel Gordon Moore, has profited all types of computing that are used by all the researchers whom all are using AI. Advance system design may have worked on a basic level, which was practiced just a couple of years back because they required computer power that was cost-restrictive or simply didn't exist. Nowadays, it is mandatory to use power for the implementation of these designs that are promptly accessible. An emotional representation is that the current age of microchips delivers 4 million times the presentation of the first single chip presented in 1971 [10].

With the rapid increase in the volume of the data by using the internet, social media, and mobile devices, it has helped in understanding the potentiality of the value of data that has led to the development of new methodologies for the analysis and management of huge data sets. In the development of AI, Big Data has played a vital role that leads to the implementation of AI methodologies in statistical models for the intellectual potentiality of data like images, text, or speech. The proposed methodologies can be enhanced or trained by making them use large data sets that are available in a ready form.

The internet and the cloud are very much similar to the concept of Big Data and are using AI techniques for two reasons: One is that data and information are available in large amounts for any computing devices that are connected through the internet and led to the implementation of AI approaches with the internet and the cloud. Another is that they have contributed in several ways in collaboration with human beings sometimes explicitly or sometimes implicitly for the training of the AI model.

New algorithms are the conventional approaches that are solving a problem or performing the task. In the past few years, several new algorithms have been developed by researchers for the improvement of the performance of machine learning, which is one of the most vital technologies for its right and empowers other technology like computer vision [11]. Most of the ML algorithms are available open-source for further enhancement that will help developers for their contributions in their research work.

1.3.3 ARTIFICIAL INTELLIGENCE AND COGNITIVE TECHNOLOGIES

The area of AI experiences both too few and several definitions. Nils Nilsson, one of the renowned researchers in this area, has defined AI as "Artificial intelligence is that activity devoted to making machines intelligent, and intelligence is that quality that enables an entity to function appropriately and with foresight in its environment" [12]. After reviewing, AI can be defined as the theory and advancement of a computer system that can be achieved by performing a task that needs human intelligence. Examples of AI are visual perception, speech recognition, translation of languages, learning, etc. [13]

We recognize the field of AI and the innovations that have been emerged from this area. The well-known press has explained AI as the upcoming of the computers as smart as – or smarter than – people. The individual innovations, e.g., comparison, are improving at performing an explicit task that human beings can only do it (Figure 1.5).

The individual technologies, conversely, are improving at performing explicit tasks that human beings can be used to have the option to do. We call these cognitive technologies (Figure 1.2), and it is these that business and public area researchers should concentrate on it.

Computer vision can be defined as the capacity of computers to recognize objects, scenes, and motions in pictures. Computer vision innovation utilizes successions of different activities of image-processing and different strategies to disintegrate the task of dissecting pictures into reasonable pieces. There are strategies for distinguishing the edges and surfaces of items in a picture, for example. Grouping strategies might be utilized to decide whether the highlights distinguished in a picture are probably going to speak to a sort of items known to the system [14]. Various applications of computer vision consist of an analysis of medical images for the improvement of the estimation, diagnosis, and proper treatment of diseases [15], recognition of face by Facebook for the identification of person automatically from the pictures [16], etc.

Machine vision is one of the related fields that is referred to the application of vision in the automation of the industry, in which the recognition of the objects is done by computer, for example, produced parts in a profoundly compelled processing plant climate – may be less difficult over the objectives of computer vision,

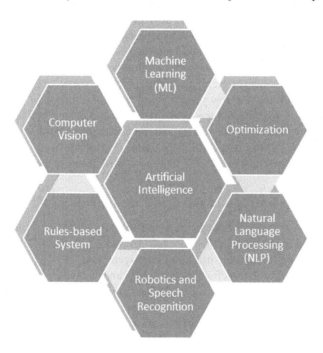

FIGURE 1.5 Artificial Intelligence different cognitive technologies.

which looks to work in unconstrained conditions. While computer vision is a field of progressing research in computer science, machine vision is a "tackled issue" – the subject not of research but rather of software engineering [17].

Machine learning is the capacity of a computer system for improving performance by the presentation of data without the need to adhere to expressly customized guidelines. At its center, machine learning is the cycle of consequently finding patterns in data. When found, the example can be utilized to make an estimation. For example, given a database of data about transactions of credit cards, date, time, merchant, location of the merchant, cost, and whether the transaction was genuine or false, ML learns designs that are estimating frauds. ML techniques assume a function in other cognitive technologies, for example, computer vision, which can prepare vision models on a huge database of pictures to improve their capacity to perceive classes of objects [18].

Natural language processing is the ability of a computer to work with a text how people do, for example, extracting significance from text or in any event, producing text into a readable, elaborately characteristic, and syntactically right. A natural language processing model doesn't comprehend text how people do, yet it can control the text in refined manners, for example, consequently recognizing the entirety of individuals and places referenced in a record; distinguishing the fundamental subject of an archive; or separating and arranging the terms and conditions in a stack of agreements, which is readable by humans. Natural language processing, similar to computer vision, contains various strategies that might be utilized together to accomplish its objectives. Language models are utilized to foresee the likelihood circulation of language articulations – the probability that a given series of characters or words is a substantial portion of a language, for example. Highlighted determination might be utilized to recognize the components of a bit of text that may recognize one sort of text from another mode of a spam email versus a real one [19].

Robotics technology incorporates computer vision, automated cognitive technologies along with high-performance sensors, actuators, and smartly designed hardware in the design of Robots that work like humans by performing a wide range of assignments in an ever-changing environment.

The main aim of speech recognition is to focus on automatic and accurate translation of human speech. There are some common problems as compared to natural language processing, which speech recognition technology has to confront, notwithstanding the troubles of adapting to different noises, differentiate among homophones ("buy" and "by" sound the equivalent), and the need to work at the speed of normal speech.

1.4 INTERNET OF THINGS

The Internet of Things (IoT) is an emerging technology having a significant socioeconomic impact. It has placed its footprint in every field of the modern world. It has paved its way from small-scale merchants to big corporate companies wherever some automation is needed. It could not be converged on a single definition as it is integrating the world of both physical and virtual. But concisely, IoT is an interconnection of physically separated objects that are commonly termed as "things" and equipped with some sensors, essential software as well as other components and technologies, which enable the connectivity and transfer of data among the inter-connected components and also over the internet. IoT is thus transforming data into facts, activities, and decisions. Here, the cellular and Wi-Fi networks play a vital role. Currently, even though it is a world of IoT, one cannot be aloof from IoT on the arrival of 5G. No new technology can indeed stand-alone without the aid of IoT and AI and blockchain technologies are currently being more connected to this emerg-ing technology. It is the IoT that turns the dumb data into meaningful information by prioritizing them and then to decision and action. Its efficiency lies in the manner how it performs the digital transformation. The following diagram shows how the IoT works in general (Figure 1.6).

1.4.1 IoT LIFECYCLE

The IoT lifecycle can be viewed as a chain of four phases, namely, collect, commu-nicate, analyze, and act. IoT can be envisioned as a large network involving inter-connected devices as well as a computing system enabled with different types of intermediary technologies [20] and as shown in Figure 1.2. The lifecycle comprises the following activities:

- **Objects tagging**: RFIDs are performing the tracing and addressing of the real-world entity.
- **Objects feeling**: The sensing equipment's primary activity is data collec-tion from all the devices existing in the surroundings.
- **Objects miniaturization**: Shrinking and nanotechnology motivated the capability of the small things for interaction and connectivity inside the "things" or "smart devices".
- **Objects thinking**: The sensors with the ability of smart intelligence inbuilt within them have a communication network with that of the internet, thereby facilitating the "things" realizing the control of the intelligence (Figure 1.7).

The IoT has brought a new perspective by extending the internet connection ahead of personal computers, mobile phones, data warehouses, grid computing, and cloud computing to the areas of agriculture, industry, power, health, transport, and pub-lic sectors. This shows the evolutionary steps of the IoT. The protocols make the

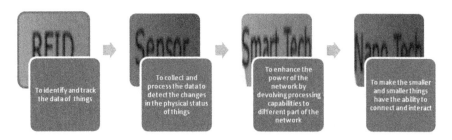

FIGURE 1.6 Functioning of IoT.

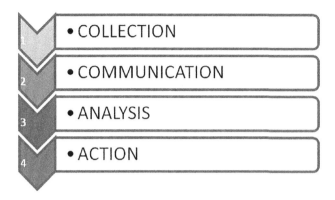

FIGURE 1.7 IoT lifecycle.

IoT system complex to some extent but the industry-oriented IoT comprises many network layers, thus making it more complex and thereby making it visualized as a system of systems. But here one needs to focus on security as complexity is the foe of security. The connection technology is more important in IoT as it should connect overall ranges of distances starting from small to very large. In wide-area transmission also, IoT is taking care of the quality.

1.4.2 IoT Framework

IoT frameworks are uncovered to their applications with the help of abstract data modeling. IoT framework is the middleware that is lying under a few application layers of IoT that provides an interface aiding the interaction for peer framework nodes. The IoT framework commonly supports many ways of information exchange [21].

1.4.2.1 Design Goals

The IoT framework has been designed over four main goals [22].

1. Reduction of time factor to enable the market arrival of the IoT solution as soon as possible
2. Reduction of noticeable deployment and functionality complexity of the IoT network
3. Making the application portable as well as interoperable
4. Making the system more reliable, ease of service, and ease of maintenance.

The IoT framework tends to hide the complex connection mechanism under the abstraction of high-level information exchange as REST. Frameworks make the IoT networks simpler by abstracting them and hiding the complex mechanisms. For developing an IoT application, build the application first with the help of any high-level language like Node.js, which can utilize framework APIs. IoT frameworks describe in detail all semantics related to all the components such as node and object [23]. This enables the developers of the IoT network to concentrate on interactivity among the nodes leaving the connection detail (Figure 1.8).

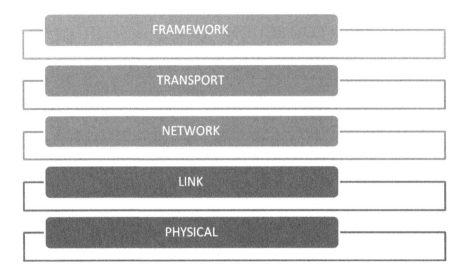

FIGURE 1.8 Layers of IoT.

Frameworks enable the applications to be more portable. Various levels in the framework help to achieve this. The system-related functions are taken care of by the bottom-most framework layer. The top layer is case-focused. For example, smart homes, smart healthcare systems, and automated process control can be included in this. With the help of the abstract of the IoT frameworks, the applications are easily built and are ready to run on the OS where one has ported the framework.

The codes of the frameworks to any of the operating systems are independent of the platform in which the applications are developed. Frameworks code migrations across platforms are possible if they are binary compatible. The device interoperability over heterogeneous surroundings is made possible by the IoT frameworks. Assume a hypothetical situation comprising many types of equipment operating various OSs as well as hardware platforms. The equipment may also have been from various platform vendors. It is expected to have multiple connection combinations and data exchange facilities for quick deployment. The IoT frameworks make it easily possible through connectivity intelligence and hide them from the application perspective.

The applications can directly act on the data with the help of the IoT frameworks, which are abstracting the layer at the application level. IoT frameworks are behaving like information-centric networks in the way they hide the complex mechanisms of the network existing in establishing the connectivity, routing mechanisms, transmitting the packets, and likewise. The devices are also abstracted by representing the physical devices logically. The frameworks interpret as though multiple nodes possessing a single IP address. Many network addresses are consolidated by the frameworks to converge into a single node. They also provide dynamic services that are built and destroyed as per message in REST. The nodes are partitioned into divisions such as domain and group. Irrespective of the various abstractions in the frameworks, the security remains. The frameworks also enable the interaction between the various nodes available in the network. Discovering, exchange of information, registering the event or activity, and notifying synchronization are some of the IoT operations. One node available in the IoT network is unable to identify similar nodes existing in the same system. It can be done through discovery only. The nodes in the other framework come to know about the supporting interfaces as well the structure of the data by inquiring to facilitate interoperation through discovery only.

1.4.3   IoT Connective Technology

IoT connectivity is defined as a connection between various systems like sensors, routers, platforms, and gateway. IoT connectivity depends on different parameters such as range, bandwidth, power consumption, and reliability [24]. Some important things should be considered at the time of choosing the right option for IoT. These things are as follows:

1. **Range**: On how much distance a device can be moved, can be installed, and can send data continuously. This distance can be a few inches or few miles, which depends on the application like an access card or drone.
2. **Battery or power consumption**: Some devices require constant power from the charging station and some devices are found at different locations or intervals. Soli sensor is an example of long battery consumption, which requires lifelong power consumption to send data; even battery replacement is also not possible at the time of data communication.
3. **Infrastructure**: Infrastructure is also an important thing to keep into consideration at the time of designing IoT.
4. **High and low bandwidth**: Bandwidth refers to the volume of data transferred over a network. Some IoT connection provides high bandwidth like Wi-Fi, and some connections allow an only small amount of data transfer.
5. **Environment**: Some IoT connections can perform perfectly in any given condition, noise, or wall but some connections cannot perform in the disturbed area. So, it is necessary to keep the environment condition at the time of designing the IoT system.

In computer science, different wireless technologies are available that allow wireless communication between devices or networks. With the help of wireless connecting technology, devices can speak to each other, which means they can transfer the data (send or receive). These connecting technologies can be divided into three categories [25].

Wi-Fi: This wireless connective technique is based on IEEE 802.11 standard (communication for wireless local area network). It also provides fast, secure, and reliable connections for peer-to-peer and personal area networks. Radio waves are used by Wi-Fi to allow communication between two devices. The connection between two hardware devices is also done with the help of Wi-Fi. In this wireless technology, a router can be connected to different objects like computers, phones, tablets, and television. This network operates in the different radio bands like 2.4,

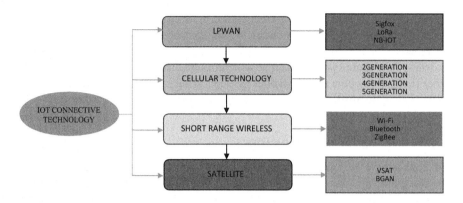

FIGURE 1.9 IOT connective technologies.

5 GHz, and possible in the third band 60 GHz with the help of an ultra-wideband channel. TCP/IP stack is used by Wi-Fi for internet connectivity [26,27] (Figure 1.9).

Bluetooth: Bluetooth comes under the wireless personal area network.

Bluetooth is another wireless technology that is used for the short-range communication. Bluetooth is based on IEEE 802. 15.1 standard. It uses lower energy and connects consumers' electronic wearable devices. IoT is not a single application where it is used but it can be used in other places where a limited communication range is not a concern (indoor asset tracking). The easy setup and low signal capabilities in crowded places make it a great choice for the application where global positioning system localization is not possible. The communication range of Bluetooth is very short like 0–10 m and uses less power consumption. The data transfer rate in Bluetooth is somewhat limited (1 Mbps). Bluetooth is the best option for embedding small smart devices like headphones, Fitbit band, smartwatches, keyboard, mouse, and speakers [28].

Bluetooth low energy (BLE) is a variant of Bluetooth and was introduced in 2006 by Nokia. BLE requires less power in comparison with Bluetooth. BLE is not perfect for phone calls but can be used for a periodic transmission of a very less amount of data. It is best suited for medical sensors like glucose meters and pulse oximeters and industrial sensors. Security is mandatory in Bluetooth mesh networking. Some security measures in Bluetooth are encryption, area isolation, security key refresh, and protection for trashcan attacks [29].

Zigbee: An open global standard is specially designed for machine to machine n/w. It is inexpensive, requires less power, and is ideal for industrial applications but can be used for home products. It is based on IEEE 802.15.4 and operates in 2.4-GHz ISM. The low latency and low duty cycle advantages of Zigbee help in maximizing the product battery life. It also provides a 128-bit advanced encryption standard. Zigbee networking technology is used in a mesh topology where nodes are connected via multiple paths. Multiple device connectivity is the ability of Zigbee, which makes it suitable for the smart home environment. In smart homes, ZigBee allows communication between various things like lights, thermostats, Heating, ventilation, and air conditioning (HVAC) controls, locks, security, and robots. The commination range of Zigbee is up to 100 m but it can further transfer data with the help of mesh network topology. The designing and maintenance of Zigbee are done by the Zigbee alliance. The latest specification of Zigbee is Zigbee 3.0, which increases flexibility for developers and ensures that all products will work through standardization at the layers of the network stack [30].

Cellular: Cellular network supports "always-on" and reliable communication via voice and video calls. The application of cellular technology in IoT is the same as consumer application, and it can be connected over 3G, 4G, and 5G networks. Broad coverage is one of the advantages of cellular. Very high operational cost, power consumption, and potential gaps are the drawbacks of cellular technology. It is not suited for the IoT application where battery-operated sensors work but it is fine for specific cases like connected care. Traditional cellular networks like 2G, 3G, and 4G use a huge amount of power and not work well where very less amount of data is transferred like smart meters, trackers, sensor (agriculture), and streets lights. It is specially designed for long-range and low-power applications. Narrowband IoT is best for infrequent communications and low bandwidth. On the other hand, long-term evolution for the machine is best for roaming and higher bandwidth application. The environmental sensors that are used to measure temperature, pressure, and wind come under the narrowband-IoT. These sensors record and send regular updates from a specific location. These devices can be used for more than 10 years if they are operated with solar power at an exact geographical location. The first generation of cellular (1G) was about only voice; the second generation supports the voice, as well as a short text message; the third generation of cellular (3G) was about voice call, text messaging, and data. The fourth generation of cellular was the same as the third generation but comparatively faster than 3G. 5G is the next generation of a cellular network with high-speed mobility that can become the future of augmented reality. Real-time monitoring for safety, e-health monitoring, and automation are some expectations with 5G [31].

Radiofrequency identification: Radio waves are used by RFID to transmit the data (small amount of data) from the RFID tag to a receiver over a short distance. The applications of RFID are retail and logistics. Real-time monitoring of products is possible only with RFID technology. An RFID tag is attached to the products so that tracking can be easy for better production planning and better stock management. Some upcoming RFID applications in retail are smart shelves, smart mirrors, and checkouts [32].

Z-wave: Z-wave is the newest type of wireless connective technology specially designed for home automation. Z-wave was developed by company Zensys in 1999, and the communication range was 100 m. In 2002, Z-waves were used by the US. The perfect example of Z-waves is Amazon echo. In the market, more than 1500 Z-waves products are available, which allow a connection between different household objects like smart doors, security alarms, remote control doors, and smart fans. Security, device compatibility, and less interference are some advantages of Z-wave over home automation [33].

Low Power Wide Area Network (LPWAN): LPWAN connective technology is designed for devices in which infrequent data transfer with low speed is required like a sensor. The gap between short-range connective techniques and cellular connective is filled by LPWANs. LPWAN supports communication from machine to machine.

LPWAN has a long battery life requirement because it covers a wide area at the time of data packet transfer using modulation technology. Line of sight is not required but some access points should be there as per area-wise. The main advantages of LPWAN are low power consumption, low cost, and long operating range (up to 10 km) [34].

Various competing technologies of LPWAN are LoRa, Sigfox, and weightless, which have different data transfer capacity and coverage levels [35].

Sigfox: It was founded in 2010 and it is a French global n/w operator that makes the connection between objects, which requires low power but needs continuous power on and transfers a small amount of data like electric smart meters and medical wearable devices [36]. In Sigfox, billions of devices are connected without establishing and maintaining the connective network. Sigfox is a cloud-based connective network where connective devices are managed on clouds. Sigfox covers the range around 10–40 km and the bandwidth is 100 HZ. Authentication and encryption are not supported by Sigfox technology [37].

LoRa WAN: Another competing technology of LPWAN is developed by Cycleo of Grenoble and acquired by Semtech. Long-range data transmission is possible with LoRa; it can cover more than 8 km in rural areas by using low power. The communication range of LoRa WAN can vary from 5 to 20 km. License-free industrial, scientific, and medical band (ISM band) is one of the advantages of LoRa WAN. It also supports advanced encryption standard of 128 bits. The frequency of LoRa WAN is different in different countries: 868 MHz (Europe), 915 MHz (North America), and 433 MHz (Asia).

Weightless: It is another NR technology that enables satellite and ground communication (uplink and downlink communication). The weightless transmission range is 2–10 km. It supports advanced encryption standard of 128 bit/256 bit. Fully acknowledged transmission with 100% is the advantage of weightlessness, which makes it different from another competing technology. The bandwidth of weightless is 12.5 kHz [38].

1.5 CONCLUSIONS

Blockchain technology provides a blazing-fast and perfectly secure environment by allowing the beneficiaries to conduct their business directly by sophisticating a decentralized way to make payments and communication. The users can generate linear and permanent indexed records. Blockchain technology allows the users to be their certificate authority with full control over their data. Blockchain technology guarantees transparency and permanence, making hacking impossible. Blockchain technology is a permanent, secure, and scalable platform on which Big Data and IoT can build. The convergence of these three technologies provides intelligent connectivity, which is having the capacity to change the style of daily lives and traditional business.

REFERENCES

[1] Svoboda, O. (2020) "Blockchains, smart contracts, decentralised autonomous organisations, and the law", *SCRIPT-ed* 17(2), 450–454.

[2] Salah, K., Rehman, M. H., Nizamuddin, N., and Al-Fuqaha, A. (2019) "Blockchain for AI: review and open research challenges", *IEEE Access* 7, 10127–10149.

[3] Zheng, P., Zheng, Z., Wu, J., and Dai, H. (2020) "XBlock-ETH: extracting and exploring blockchain data from ethereum", *IEEE Open J Comput Soc* 1, 95–106.

[4] Yaga, D., Mell, P., Roby, N., and Scarfone, K. (2019) "Blockchain technology overview".

[5] Kumar, H., Agrawal, K., Manu, M. R., Indrakumari, R., and Balamurugan, B. (2020) "Blockchain use cases in big data", In *Blockchain, Big Data, and Machine Learning: Trends and Applications*, p. 111, CRC Press, United States.

[6] Indrakumari, R., Poongodi, T., Saini, K., and Balamurugan, B. "Consensus algorithms–a survey", In *Blockchain Technology and Applications*, pp. 65–78, CRC Press, United States.

[7] Zheng, Z., Xie, S., Dai, H., Chen, X., and Wang, H. (2017) "An overview of blockchain technology: architecture, consensus, and future trends", In *2017 IEEE International Congress on Big Data (BigData Congress)*, pp. 557–564, Honolulu, HI. doi: 10.1109/BigDataCongress.2017.85.

[8] Anand, M. V., Poongodi, T., and Saini, K. (2020) "Bitcoins and crimes." In *Blockchain Technology and Applications*, pp. 223–245, CRC Press..

[9] CB Insights Data, Deloitte Analysis. The $2 billion figure includes investments in companies selling AI technology or products with the technology embedded.

[10] Nilsson, N. (2010) *The Quest for Artificial Intelligence*, p. 13, Cambridge: Cambridge University Press.

[11] Oxford Dictionaries, "Definition of artificial intelligence," http://www.oxforddictionaries.com/us/definition/american_english/artificialintelligence, accessed October 3, 2014.

[12] Russell and Norvig, Artificial Intelligence.

[13] Chen, C. H. (2014) *Computer Vision in Medical Imaging*, Singapore: World Scientific Publishing Company.

[14] J. Mitchell, "Making photo tagging easier," Facebook, https://www.facebook.com/notes/facebook/makingphoto-tagging-easier/467145887130, accessed on October 18, 2014.

[15] Graves, M., and Batchelor, B. G. (2003) *Machine Vision for the Inspection of Natural Products*, p. 8, London: Springer.

[16] For instance, Microsoft recently announced that it had developed a computer vision system able to identify dog breeds. It relies in part on machine learning techniques and was trained using a database of millions of images. See Microsoft Research, "On Welsh Corgis, computer vision, and the power of deep learning," http://research.microsoft. com/en-us/news/features/dnnvision-071414. aspx?0hp=002c, accessed October 6, 2014.

[17] Russell, S. J., and Norvig, P. (1995) *Artificial Intelligence*, pp. 860–885, Englewood Cliffs, NJ: Prentice Hall.

[18] Danowitz, A., et al., (2014) "CPU DB: recording microprocessor history", *ACMQueue* 10(4), http://queue.acm.org/detail. cfm?id=2181798, accessed October 11, 2014.

[19] Multiple researchers have devised algorithms that have improved the performance of machine learning. Google Scholar finds some 500,000 scholarly papers on the topic of neural networks, for example, published since 2006. Geoffrey Hinton is a widely published and cited researcher in this area credited with several important innovations. See Geoffrey Hinton, "Home Page of Geoffrey Hinton," http://www.cs.toronto.edu/~hinton/, accessed October 6, 2014. Other researchers who are widely recognized for contributions in this area include Yann LeCun (See Yann LeCunn, "Yann LeCun's Home Page," http://yann. lecun.com/, accessed October 9, 2014), and Yoshua Bengio (see Yoshua Bengio, "Yoshua Bengio's Research," http://www.iro.umontreal. ca/~bengioy/yoshua_en/research.html, accessed October 9, 2014). Recently, Microsoft demonstrated a new machine learning architecture that dramatically accelerates the machine learning process, improving precision and accuracy. See, Microsoft Research, "On Welsh Corgis, computer vision, and the power of deep learning," http://research.microsoft. com/en-us/news/features/dnnvision-071414.aspx?0hp=002c, accessed October 6, 2014.

[20] Mani, V. (2017) "A view of blockchain technology from the information security radar", *ISACA* 4.

[21] Zheng, Z., et al., (2017) "An overview of blockchain technology: architecture, consensus, and future trends", In *2017 IEEE International Congress on Bigdata*, pp. 557–564, IEEE.

[22] Cheruvu, S., Kumar, A., Smith, N., and Wheeler, D. M. (2012) "Connectivity technologies for IoT", In *Demystifying Internet of Things Security*, pp. 347–411, Berkeley, CA: Apress.

[23] https://www.digiteum.com/internet-of-things-connectivity-technologies.

[24] https://iot-analytics.com/iot-segments/iot-connectivity/.

[25] https://www.link-labs.com/blog/types-of-wireless-technology.

[26] https://www.iotforall.com/wifi-role-iot.

[27] https://www.digiteum.com/internet-of-things-connectivity-technologies#: ~:text=Bluetooth%20is%20another%20short%2Drange,to%20continuously%20send% 20status%20data.

[28] https://www.aeris.com/news/post/bluetooth-for-iot/.

[29] https://bliley.com/wireless-technologies-for-iot.

[30] Jia, X., Feng, Q., Fan, T., and Lei, Q. (2012) "RFID technology and its applications in the internet of things", In *2012 2nd International Conference on Consumer Electronics, Communications and Networks (CECNet)*, pp. 1282–1285, IEEE.

[31] https://www.safety.com/z-wave/.

[32] https://www.the-ambient.com/guides/zwave-z-wave-smart-home-guide-281.

[33] https://en.wikipedia.org/wiki/LPWAN.

[34] https://blog.antenova.com/comparing-low-power-wide-area-networks-for-iot.

[35] https://en.wikipedia.org/wiki/Sigfox.

[36] https://www.sigfox.com/en/what-sigfox/technology.

[37] https://www.avnet.com/wps/portal/apac/resources/article/sigfox-is-the-worlds-leading-provider-of-connectivity-for-the-internet-of-things/.

[38] https://www.gsm-modem.de/M2M/iot-university/nb-iot-power-consumption/.

2 A Decentralized Privacy-Preserving Blockchain for IoT and Big Data in Healthcare Applications

L. Godlin Atlas and K.P. Arjun
Galgotais University

Bindu Babu
Easwari Engineering College

CONTENTS

2.1   INTRODUCTION

With the Internet of Things (IoT) and applications, for instance, wearable individual prosperity trackers, medical care has gotten one of the most crucial bits of human continues with, provoking an emotional augmentation in clinical immense data. To streamline the end and treatment measure, human administration specialists are right now grasping IoT-based wearable development. For years have seen a lot of sensors, devices, and vehicles that are related through the Internet. One type of innovation – far away from patient checking – is fundamental these days for the care and treatment of patients. What's more, each restorative administration relationship

DOI: 10.1201/9781003081180-2

over the thought continuum is starting at now running to expand an edge over their companions by using gigantic data examination to offer the best response for the puzzle of the Triple Aim. Man-made thinking and blockchain have promptly become the instruments of choice for architects, providers, and payers planning to reinforce their prosperity IT establishment with innovative, amazing data the leader's capacities. Be that as it may, the improvement provides grave insurance and security stresses over the information move and the records of data trades. These security and insurance issues of clinical information could lead to a deferral in the treatment process, regardless of endangering the life of the patient. The issues of insurance and security are essential to the prosperity of our therapeutic administration data. The conceivable furthest reaches of blockchain, in our human organizations industry given the various players, for example, confirmation, trained professionals, medications, specialists or even more all the patient where exchanges, for example, clinical idea records ; quiet starter results may be executed forever either on an open, on private, or on cross-variety record. In this chapter, a novel security framework is proposed for social protection intuitive media data through blockchain strategy by creating the hash of each data with the goal that variations in the data or dividing of prescriptions may be seen in whole blockchain sort-out clients. Additional insurance and security enhancement properties in our model rely upon forefront cryptographic natives. The game plans mentioned here make the applications of IoT data and trades more secure and mysterious over a blockchain-based framework (Figure 2.1).

The short development of human offerings treatment techniques and strategies might also be the reason for diverse correspondence and capability issues among various sellers (medical doctors, health safety companions, drug specialists) and sufferers. Several issues are checked out by means of the patients in the scientific parts and their related administrations in which several big private emergency clinics and the professionals utilize an effectively lifeless patient to get cash from his/her own family through maintaining them in a misguided judgment that the affected person

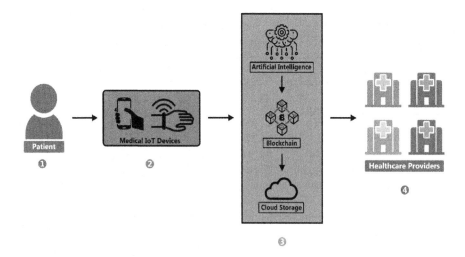

FIGURE 2.1   Application of IoT with AI and blockchain.

is alive. In a provided to ensure about commission, experts extra consistently than at this point don't advocate pills which are to be had in the crisis facilities pharma-shop regardless of the openness of significantly less costly restorative medications in the external market. Sometimes, to make them their benefit, even though the clinical centers related to pathology laboratories won't accept notoriety, the specialists will be intimating the patients to do the laboratory tests or purchase the meds from that association only. In extension, though the entire data of patients are taken care of on the web; nevertheless, after the exchanging of subject matter experts, patients save no choices to get to or redo their records. After exchanging, the patient is necessary to follow the entire test again with the new trained professional and need to pay twice. In a sales to manufacture security and trust among different parts, there is a need to propose some immediate methods of reasoning for mounting the going with accessory of IoT deals with working on the security of patient's data, straightforwardness acknowledges a fundamental work.

A couple of analysts, makers, and experts have previously presented reviews on a couple of utilization cases of therapeutic contributions in IoT and blockchain into significantly few regions for adapting to the media information. In this section, the factor utilizing point portrayal of the IoT in friendly protection, blockchain capacities, and its work in human administrations has been explained. To securely cross the vast amount of real-time information from IoT gadgets, Yu et al. [33] have proposed a blockchain degree that gainfully hits contemporary realities effortlessly. In this paper, to design an economic regard statistics pass, the makers have prepared a blockchain level using specific advancements, for example, middle factor arranging in sharp gadgets, the coursed cartoon of the nearby vicinity, and a know-how computation to understand the decentralized technique of shrewd gadgets. Similarly, in a solicitation to accumulate and supervise top-notch proportions of information from IoT devices and to decide the mining and capability issues from extraordinary IoT companies, Ozyilmaz et al. [25] have proposed a blockchain construction to handle those issues. On 13 November 2017, the Food and Drug Administration (FDA) validated the primary pill with the sensor inner it (aripiprazole drugs with sensor) that could consent to if an animated individual has gulped it. This present tablet's sensor sends messages to a wearable repair, and the real restore conveys the message to a cell software at the phone telephone. This development needs to be a first-rate addition for endless illness and passionate prosperity troubles.

In this bankruptcy, we positioned out some of our essential important motivations to take advantage of the blockchain in healthcare from Ref. [3–8], the region the cited authors have stated the cutting-edge traits within the studies of blockchain. Creation to bitcoin is cited right here [1]; the chances are endless of ways the hidden innovative might be utilized in selective methods outside of the financial domain. Various attempts at utilizing the improvement of blockchain age identified with the financial domain [2–7]. To consider the materialness of this far achieving mechanical specifically in medical services, shrewd towns, and IoT. Many protection breaches or facts losses are already keeping medical data [8-9]. Hackers are seeking treasured health facts wherein it is able to include pertinent records for identification theft. Medical record possession is any other key element when discussing fitness records. Numerous varieties of information will come as many distinctive patterns.

The patterns are referred to right here, which tells the type of statistics within the form of photographs, videos, reviews, and many others. The records are also available in exquisite codecs counting on the systems in use via the way of the provider referred to. The records integrity turns into dominant and the requirement desires to be a herbal event in the vicinity to make certain that the integrity of the information is maintained in order to make sure the information have no longer been changed, destroyed, or eliminated. The entrance the realities should be overseen through the patients; skill, they should now not be in a job to change it both. The affected person's data needs to be regular and reachable throughout institutional obstacles [10-14].

Further, for tending to the protection and safety trouble in IoT gadgets, Yong Yu et al. [15-18] have built up an IoT structure incorporated with blockchain innovation that guarantees quite several promises, the affirmation of facts move, appealing versatility with validation and decentralized system for the installments, which are the unique offices furnished with the blockchain-empowered IoT framework. To approve the proposed marvel, the authors have advocated a few arrangements utilizing Ethereum programming for demonstrating the contribution of blockchain innovation in IoT. Further, writing writers have proposed special IoT solutions utilizing blockchain innovation in one-of-a-kind fields.

In Ref. [19-20], there used to be a prologue to techniques for utilizing blockchain to deliver evidence of prespecified endpoints in clinical preliminaries. Irving and Holden tentatively endeavored this sort of procedure the utilization of a precise introductory show the spot result trading had at present been represented . They confirmed the usage of blockchain as a minimal effort, freely verifiable approach to evaluate and confirmed their liability of scientific considers. We utilize lightweight superior mark plans in our model influenced with the aid of Ref. [21]. We've got seen some solid work within the placing of smart urban areas and Big Data, as of overdue by using Wu and Ota in Refs. [22–23]. They have in reality had the choice to pay attention to how the smart cities together with IoT are the safe and green transmission of clinical records.

The authors have contributed some paintings as of now in cryptanalysis of ARX figures and other security calculations [24–30]. Dey et al. [11] have proposed square chain engineering to defeat the postponement figuring out with the presenting affected person's report or healing. The creators have reduced this trouble with a biosensor dimension that collects and shops consistent statistics of sufferers in the square chain. Similarly, keen agreements were deployed to parent the bills with protection companions. Ngamsuriyaroj et al. [31] have proposed an impenetrable package deal conveyance element using blockchain innovation that improves the integrity and secrecy of the two data and customers. The proposed device assessed addition and study-compose operations. Until now, writers have proposed special-use instances of rectangular chain innovation. Ability, scarcely any examiners have introduced the thought of rectangular chain in IoT-human administrations. The greater phase of the researchers has checked their consequences concerning take a look at their cryptographic parameters. The quantity of this paper is to offer a shape of IoT social insurance plan the use of rectangular chain technological and dissect the effects regarding pernicious IoT gadgets, probabilistic situations of validating the purchaser's solicitation and clients demand time-making use of square chain hubs.

2.2   SECURITY PITFALLS IN EXISTING SYSTEM

The precept fear in healthcare frameworks is the safe and efficient transmission of the clinical data. Healthcare statistics is a lucrative objective for programmers, and alongside these traces, making sure about ensured well-being facts is the imperative notion of human offerings suppliers. Medicinal services have emerged as the fundamental goal for cybercriminals. For instance, digital assaults on scientific devices or health facts have grown to be extra common in the latest decade. In any case, the powerlessness to erase or trade statistics from squares makes blockchain innovation a fantastic innovation for the human services framework and may want to forestall these issues. However, blockchain innovation in its special shape isn't always a sufficient arrangement. In this area, challenges appeared through blockchain and IoT designs and answers for taking care of these issues.

- To warrant heartiness and adaptability and to dispense with many-to-one traffic flows, we need a decentralized framework. Implementing decentralized frameworks will help to maintain the problems related to record defer, which can be stopped.
- Client's PC or cloud organizations keep unpreserved records that desire to be moved to blockchain frameworks. During transmission, the data should be modified or lost. The conservation of such off-base altered information expands the weight to the framework and can purpose the loss of the patient (demise). In this way, to assure that records aren't always modified, a lightweight computerized signature [32] plot is utilized. On the beneficiary side, information is verified with the customer's advanced mark, and at whatever point gained successfully, it sends a receipt of measurements to the patient.
- Solving PoW is computationally concentrated; be that as it may, IoT gadgets are helpful to asset limited. Moreover, the IoT mastermind consolidates various centers, and blockchain scales ineffectively as the range of hubs in the system increments. In the proposed model, arrange is isolated into a few corporations instead of a solitary chain of blocks, and consequently, a solitary blockchain is not in charge of all hubs. Rather, the hubs are unfolding more than a few bunches.
- Storing IoT big facts over the blockchain isn't always pragmatic, and alongside these strains, cloud workers are utilized to save encoded information squares. The data is shielded over the cloud because of extra cryptographic security like the unrivaled signature and particular necessity encryptions. Regardless, it might reason inconvenience about trusted in outcasts. For this reason, all exchanges are put away in various squares and make a joined hash of each square using MerkleTree.
- Medical gadgets or health statistics should be particular and cannot be changed by the way of hackers. To save the records from developers, we are using a twofold encryption plot. Here, the twofold encryption plot would not insinuate scrambling comparative information using two keys yet then again encryption of the data and again encryption of key, which was used to encode data. We encode the realities of the utilization of lightweight ARX

figuring's and a while later scramble the key utilizing the open key of the beneficiary. Also, the Diffie–Hellman key exchange technique is utilized to go the open keys, and consequently, getting the keys is practically incomprehensible for an aggressor.

2.3   PROPOSED SYSTEM AND IMPLEMENTATION

In preferred strategies, the reasonability and the main organization of patient's reviews, documentation, and examinations or prescription meds created by means of specialists were regulated freely. In solicitation to make blessings by way of the patients, professionals may also recommend inconsequential assessments and meds to buy from unequivocal stores [33-34]. Further, the affected person is completely uninformed of these medical reports and experts don't supply the entire documentation to the patients wherein the affected person is satisfied to switch his/her vital consideration physician. The criminal operations or advantages made with the patient are uninformed by using him/her. A few super methodologies, as an instance, sharp devices, had been used where the patient's file and treatment are completed through machines or insightful gadgets. The all-out method of affected person's treatment as an example from diagnosing to reestablishing or invoice quantities via insurance relates, the total document, or gadget has been positioned over the cloud. In addition, throughout the proposal cycle in which the professional implies his/her affected person to every other educated expert, the patient's scientific document has been moved on line where the treatment is executed over wonderful objects. Regardless, diverse problems should be tended to with those frameworks; for instance, a further security affiliation may append with positive disaster centers and undertaking to accumulate the benefits from the patients. Sometimes for a condition wherein drug associations put together the meds from tough substances, intermediates can also gain the aspect deftly.

In everyday structures, the reasonability and the chiefs of patient's reviews, documentation, and assessments or prescription medications delivered with the aid of experts have been administered autonomously. In solicitation to make benefits with the aid of the sufferers, specialists might also endorse trivial assessments and medicinal drugs to purchase from unequivocal stores [35]. In addition, the affected person is absolutely unconscious of these scientific reviews and specialists don't give the all-out documentation to the patients wherein the affected person is demanding to switch his/her crucial attention physician. The criminal operations or benefits made with the patient are oblivious by means of him/her. Some exquisite methodologies, as an instance, short gadgets, have been used wherein the patient's record and remedy are accomplished via machines or extremely good things. The absolute strategy of patient's remedy, for example, from diagnosing to reestablishing or invoice quantities through insurance relates, the total record or machine has been positioned over the cloud. In addition, in the course of their honor cycle in which the professional shows his/her affected person to any other skilled professional, the affected person's scientific file has been moved on-line wherein the treatment is performed over eager articles. Anyways, various troubles ought to be tended to with these frameworks; for example, a further safety affiliation may additionally connect with certain clinical facilities and undertaking to achieve the benefits from the sufferers. Every now and then for a situation wherein drug institutions plan the medicines from unrefined materials, intermediates may also gain the element results easily.

The device plan of the proposed social safety shape is addressed here that calls for an internet software containing terminations, for instance, a the front cease that receives together with sufferers and a again give-up, which offers inside correspondence using square chain. The unique interest goes likely as an affiliation among these completions. The proposed social assurance shape is modest by utilizing introducing electronic correspondence some of the victims and shippers. At some stage in the execution of the returned cease where the blockchain correspondence degree happens, there are the substances that can be associated all collectively of center factors. To understand the medical care blockchain measure, we've used forms of center points, for instance, earthmovers or checking hubs and other are executing centers. The digger's mission is to advise the proper or incorrect trade to breed the records base popularity upon its devotion or excusal. In any case, the executing middle factor's venture is to examine whether the alternate accumulating within the tractors is legitimate or no longer. Within the occasion that the fee is full-size, it'll furthermore gain the digger's regard and execute it into the square. Here, we've got offered two digital machines (VM) structures as shown in Figure 2.7 to impersonate a veritable blockchain device. Gadget 1 has the executing center, while system 2 executes the two diggers and executing centers.

Rather than handiest one sort of encryption strategy, we use both encryption plans, mainly Symmetric and uneven for special purposes. Symmetric figuring (non-public key encryption) can be used for the non-public key or symmetric key in our counts, and a comparable key can be used for encryption and decoding on each aspect of the transmission (Figure 2.2).

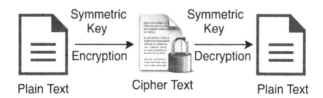

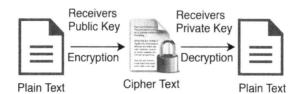

FIGURE 2.2 Asymmetric encryption.

The precept fear in healthcare frameworks is the safe and efficient transmission of the clinical data. Healthcare statistics is a lucrative objective for programmers and alongside these traces, making sure about ensured well-being facts is the imperative notion of human offerings suppliers. Medicinal services have emerged as the fundamental goal for cybercriminals. For instance, digital assaults on scientific devices or health facts have grown to be extra common in the most latest decade. In any case, the powerlessness to erase or trade statistics from squares makes blockchain innovation a fantastic innovation for the human services framework and may want to forestall these issues. However, blockchain innovation in its special shape isn't always a sufficient arrangement. In this area, challenges appeared through blockchain and IoT designs and answers for taking care of these issues.

- To warrant heartiness and adaptability and to dispense with many-to-one traffic flows, we need a decentralized framework. Utilizing such decentralized frameworks, the single reason of disappointment or records defer problems can be killed.
- Client's PC or cloud organizations keep unpreserved records that desire to be moved to blockchain frameworks. During transmission, the information ought to be modified or lost. The conservation of such misinformed adjusted information develops the weight to the structure and can reason the deficiency of the patient (defeat). Thus, to the affirmation that records aren't for the most part modified, a lightweight electronic imprint [36] plot is utilized. On the recipient side, data is verified with the client's serious imprint, and at whatever point picked up enough, it sends a receipt of estimations to the patient.

Clinical devices or well-being measurements ought to be specific and can't be changed by the method of programmers. To save the records from developers, we are using a twofold encryption plot. Here, the twofold encryption plot would not recommend scrambling near data utilizing two keys, yet, on the other hand, encryption of the information and again encryption of key were utilized to encode information. We encode the realities of the utilization of lightweight ARX estimations, and a while later, we scramble the key utilizing the open key of the beneficiary. Additionally, the Diffie–Hellman key trade strategy is used to go the open keys, and subsequently getting the keys is unimaginable for an attacker.

- Clinical devices or well-being measurements ought to be specific and can't be changed by the method of programmers. To save the records from developers, we are using a twofold encryption plot. Here, twofold encryption plot would not propose scrambling relative data utilizing two keys, yet, on the other hand, encryption of the information and again encryption of key were utilized to encode information. We encode the realities of the utilization of lightweight ARX estimations, and a while later, we scramble the key utilizing the open key of the beneficiary. Additionally, the Diffie–Hellman key trade strategy is used to go the open keys, and subsequently getting the keys is unimaginable for an attacker.

- Medical gadgets or health statistics should be particular and cannot be changed by way of hackers. To spare the records from programmers, we are utilizing a twofold encryption plot. Here, the twofold encryption plot would not allude to scrambling similar data utilizing two keys yet alternatively encryption of the information and once more encryption of key were utilized to encode information. We encode the facts with the use of lightweight ARX calculations and afterward scramble the key using the open key of the recipient. Also, the Diffie–Hellman key exchange technique is utilized to go the open keys, and consequently getting the keys is practically incomprehensible for an aggressor.

2.4   PERFORMANCE EVALUATION

The precept worry in healthcare frameworks is the safe and efficient transmission of the scientific data. Healthcare records are a rewarding objective for programmers and along these strains making certain about ensured wellness data is the quintessential thought of human services suppliers. Medicinal services have come to be the indispensable goal for cybercriminals [37]. For instance, digital assaults on scientific devices or wellness information have grown to be extra common in the most current decade. In any case, the powerlessness to erase or change facts from squares makes blockchain innovation the nice innovation for the human services framework and may want to forestall these issues. However, blockchain innovation in its unique shape isn't always an adequate arrangement. In this area, challenges are regarded through blockchain and IoT designs and solutions for taking care of these issues.

- To warrant heartiness and adaptableness and to dispense with many-to-one traffic flows, we need a decentralized framework. Utilizing such decentralized frameworks, the single purpose of sadness or statistics defer issues can be killed.
- On the recipient side, information is verified with the consumer's digital signature, and on every occasion acquired successfully, it sends a receipt of information to the affected person. Consumer's PC or cloud agencies store unpreserved insights that need to be moved to blockchain frameworks. At some point of transmission, the facts may be altered or misplaced. The preservation of such the weight to the shape might cause the deficiency of the patient (destruction). Along these follows, to guarantee that data isn't modified, a lightweight mechanized imprint [38–44].
- Comprehending PoW is computationally notion; in any case, IoT devices are help to be restrained. In a manner, the IoT plan conveys numerous facilities, and blockchain scales inadequately as the number of middle points in the device will increase. Inside the proposed version, set-up is remoted into a couple of associations alternatively than a single chain of squares, and thus, a singular blockchain isn't commonly reliable for all middle points. Or maybe, the middle factors are unfurled more noteworthy than more than one bundles Placing away IoT huge measurements over the blockchain isn't always businesslike, and along these strains, cloud employees are used to

keep encoded insights squares. The insights are shielded over the cloud because of greater cryptographic protection just like the serious signature and special prerequisite encryptions. Notwithstanding, it may reason a trouble around put stock in untouchables. Consequently, all trades are looked after in exclusive squares and make a joined hash of every rectangular creation usage of Merkle tree [44–48].

- Medical devices or well-being records have to be terrific and cannot be modified by means of the approach of programmers. To spare the records from engineers, we are utilizing a twofold encryption plot. Here, the twofold encryption plot does not advocate scrambling nearly same information using two keys but instead encryption of the facts and again encryption of key were used to encode facts. We encode the data with the use of mild-weight ARX computations and in some time scramble the key introduction usage of the open key of the beneficiary. Additionally, the Diffie–Hellman key alternative framework is used to go the open keys, and therefore, getting the keys is essentially unfathomable for an attacker.

2.4.1 Mining Attack

Recollect an adversary hacked some social events takes and commenced controlling unmistakable pack heads. In this type of circumstance, fake mining is feasible anyway as soon as it's far recognized through different bunch heads or center points; they could anticipate a totally first-rate stretch to have a look at the faux amassing heads. This is due to the fact in our mannequin in the tournament that any pack head favors a square; at that point, it will comprise a mechanized mark over that rectangular, and besides the serious imprint, different gathering heads might not apprehend another square in the framework. At the point when a phony group head is recognized through the framework, it will in universal be modified by the centers in that bundle.

2.4.2 End Point Security Attacks

Endpoints are where men and women interface with the framework. In traditional conditions, endpoints are below the control of the challenge although in a blockchain environment give up purchasers and diggers make use of their frameworks to get the right of entry to the administrations of blockchain. To get to the blockchain, one wants to have a private key helpful. Programmers take client's keys by several means, for example, assaulting their e-mail money owed or laptops.

2.4.3 Sybil Attacks

This is an assault where the programmer imagines the same quantity of hubs in the blockchain simultaneously. At the factor when the programmer takes the cost of the widespread level of companion hubs in the blockchain network, each change receives acknowledged as the lousy actor owns a large part of the hubs. It is enormously tough to do a Sybil assault on a greater prolonged blockchain system; for example, bitcoin as large computing strength is required.

2.4.4 ROUTING ATTACKS

Steering assault fundamentally incorporates two separate attacks. The initial one is "Apportioning assault" the spot the malignant web organization supplier (ISP) allocations the blockchain local area into something like two get-togethers by utilizing seizing now relatively few important organizations centers. The subsequent boost is "Defer Attack" in which the vindictive ISP defers the rectangular engendering that makes the blockchain organize defenseless to twofold spending.

The probabilistic situations of the affirmation segment, the spot in the wake of fostering the volume of MNs (for even the diggers or pal focus focuses) with the advancement in creating size, the two philosophies recognize the authentic center point. The proposed structure that continues up a blockchain among each middle point can identify the trusted in core point. The precision will be, moreover, improving with time on account of the expulsion of perceived MNs from the structure. Recognizable proof of MNs subject to have faith and removal of diagnosed MNs did not block the exhibition of exclusive hubs. The proposed part calculates the have confidence and ranking of their middle factors after a particular period. The middle points that are sabotaged and act maliciously will have low appraising and belief (because of much less aspect drop extent, wormhole, and debasement attack) and ought to by no means be considered in the future.

Further, the exactness of the proposed approach was once multiplied with the time because of the ejection of special MNs from the structure. The area of MNs depends upon belief where the departure of perceived MNs does no longer stop the introduction of a range of core points. The proposed segment measures the belief of various middle factors after every precise timeframe the place centers that are sabotaged and elevate threateningly will have low assessing and have faith because of excessive factor mishap extent, darkish opening, and distortion assault and would in no way be viewed in the future.

2.5 CONCLUSION

Healthcare corporations are turning up new technology to guide healthcare vendors and sufferers. The purpose is to offer control for patients over their fitness statistics. Healthcare data throughout healthcare establishments are complicated, which probably will increase the studies and clinical electiveness. Due to its decentralized nature, it gives new borderless incorporated healthcare services to patients. It also is of the same opinion the healthcare providers to record and manipulate the clinical transactions via a network with none imperative authority. But the effectiveness of information sharing between exclusive carriers and hospital structures is confined via the interoperability venture. This loss of coordination in statistics management remains a substantial barrier in its implementation.

As the scientific records are diverse in nature, they obtain care from multiple institutions. Furthermore, a new case has a look at of the blockchain primarily based healthcare system is presented along with its perils and pitfalls within the current layout structure. The mitigations are furnished to solve the barrier between blockchain transparency and information confidentiality by the crowdsourcing method.

REFERENCES

[1] Mohammed, M., Khan, M. B., and Bashier, E. B. M. (2017) *Machine Learning Algorithms And Applications*, Milton Park: Taylor & Francis Group, LLC.

[2] Bohra, H., Arora, A., Gaikwad, P., Bhand, R., and Patil, M. R. (2017) "Health prediction and medical diagnosis using Naive Bayes", *Int J Adv Res Comput Commun Eng* 6(4), 32–35.

[3] Maithili, A., Vasantha Kumari, R., and Rajamanickam, S. *Int J Mod Eng Res (IJMER)* 1(1), 57–64.

[4] Wang, S., and Cai, Y. (2018) "Identification of the functional alteration signatures across different cancer types with support vector machine and feature analysis", *Biochim Biophys Acta Mol Basis Dis* 1864(6 Pt B), 2218–2227.

[5] Huang, S., Cai, N., Pacheco, P. P., Narandes, S., Wang, Y., and Xu, W. (2018) "Applications of support vector machine (SVM) learning in cancer genomics", *Cancer Genomics Proteomics* 15(1), 41–51.

[6] Ogbuabor, G., and Ugwoke, F. N. (2018) "Clustering algorithm for a healthcare dataset using silhouette score value", *Int J Comput Sci Inf Technol (IJCSIT)* 10(2), 27–37.

[7] Han, J., Kamber, M., Pei, P. (2012) "10 - Cluster Analysis: Basic Concepts and Methods", In *The Morgan Kaufmann Series in Data Management Systems, Data Mining* (Third Edition) (J. Han, M. Kamber, and J. Pei, Eds.), pp. 443–495, Morgan Kaufmann.

[8] Kabeshova, A., and Reesfrance, D. (2016) Decision Analysis in Public Health, EHESP.

[9] Zheng, B., Yoon, S. W., and Lam, S. S. (2014) "Breast cancer diagnosis based on feature extraction using a hybrid of k-means and support vector machine algorithms". *Expert Syst Appl* 41(4), 1476–1482.

[10] Khare, A., Jeon, M., Sethi, I. K., and Xu, B. (2017) "Machine learning theory and applications for healthcare", *J Healthcare Eng* 2017, 2. https://doi.org/10.1155/2017/5263570.

[11] Alam, S., Kwon, G.-R., Kim, J.-I., and Park, C.-S. (2017) "Twin SVM-based classification of Alzheimer's disease using complex dual-tree wavelet principal coefficients and LDA", *J Healthcare Eng.*

[12] van Hartskamp, M., Consoli, S., Verhaegh, W., Petkovic, M., and van de Stolpe, A., (2019) "Artificial intelligence in clinical health care applications", *Interact J Med Res* 8(2), e12100.

[13] Byun, H.-S., Hwang, H., and Kim, G.-D., (2020) "Crying therapy intervention for breast cancer survivors: development and effects", *Int J Environ Res Public Health* 17(13).

[14] Lowe, D. G., and Muja, M. (2014) "Scalable nearest neighbor algorithms for high dimensional data", *IEEE Trans Pattern Anal Mach Intell* (cited 324 times, HIC: 11, CV: 69).

[15] Chandrashekar, G., and Sahin, F. "A survey on feature selection methods", *Int J Comput Electr Eng* (cited 279 times, HIC: 1, CV: 58).

[16] Corchado, E., Graña, M., and Wozniak, M. (2014) "A survey of multiple classifier systems as hybrid systems", *Inf Fusion* 16, 3–17. (cited 269 times, HIC: 1, CV: 22).

[17] Locatello, F., Bauer, S., Lucic, M., Ratsch, G., Gelly, S., Scholkopf, B., and Bachem, O. (2019) Challenging common assumptions in the unsupervised learning of disentangled representations, In *International Conference on Machine Learning*, pp. 4114–4124.

[18] Sharma, K., Rafiqui, F., Attri, P., and Yadav, S. K. (2019) "A two-tier security solution for storing data across public cloud", *Recent Patents Comput Sci* 12(3), 191–201.

[19] Sharma, K., and Shrivastava, G. (2014) "Public key infrastructure and trust of web-based knowledge discovery", *Int J Eng Sci Manage* 4(1), 56–60.

[20] Shrivastava, G., Kumar, P., Gupta, B. B., Bala, S., and Dey, N. (Eds.) (2018) *Handbook of Research on Network Forensics and Analysis Techniques*, Hershey, PA: IGI Global.

[21] Ahmad, F. A., Kumar, P., Shrivastava, G., and Bouhlel, M. S. (2018) "Bitcoin: digital decentralized cryptocurrency", In *Handbook of Research on Network Forensics and Analysis Techniques*, pp. 395–415, Hoboken, NJ: IGI Global.

[22] Srivastava, S. R., Dube, S., Shrivastaya, G., and Sharma, K. (2019) "Smartphone triggered security challenges-issues, case studies, and prevention", In *Cyber Security in Parallel and Distributed Computing: Concepts, Techniques, Applications and Case Studies*, Hoboken, NJ: Oreilly, pp. 187–206.

[23] Amit, R., and Zott, C. C. (2015) "Crafting business architecture: the antecedents of business model design", *Strateg Entrep J* 9(4), 331–350.

[24] Stevenson, T. (2002) "Anticipatory action learning: conversations about the future", *Futures* 34(5), 417–425.

[25] Inayatullah, S. (2006) "Anticipatory action learning: theory and practice", *Futures* 38(6), 656–666.

[26] Coghlan, D., and Brannick, T. (2010) *Doing Action Research in Your Own Organization*, 3rd ed., London, UK: Sage.

[27] Reason, P. (2006) "Choice and quality in action research practice", *J Manage Inquiry* 15(2), 187–202.

[28] Tsoukas, H., and Shepherd, J. (Eds.) (2004) *Managing the Future: Foresight in the Knowledge Economy*, Hoboken, NJ: Blackwell Publishing Ltd.

[29] Ramos, J. M. (2006) "Dimensions in the confluence of futures studies and action research", *Futures* 38(6), 642–655,

[30] Floyd, J. (2012) "Action research and integral futures studies: a path to embodied foresight", *Foresight* 44(10), 870–882,

[31] Bryman, A., and Bell, E. (2011) *Business Research Methods*, 3rd ed., New York, NY: Oxford University Press.

[32] Yin, R. K. (2009) *Case Study Research: Design and Methods*, 4th ed., Newbury Park, CA: Sage.

[33] Nakamoto, S. (2008) *Bitcoin: A Peer-to-Peer Electronic Cash System* [Online]. Available: https://bitcoin.org/bitcoin.pdf.

[34] Yli-Huumo, J., Ko, D., Choi, S., Parka, S., and Smolander, K. (2016) "Where is current research on blockchain technology?—a systematic review", *PLoS One* 11(10), e0163477.

[35] Vukolić, M. (2016) "The quest for scalable blockchain fabric: proof-of-work vs. BFT replication", In *Open Problems in Network Security*, vol. 9591 (J. Camenisch and D. Kesdoğan, Eds.), Cham, Switzerland: Springer.

[36] Christidis, K., and Devetsikiotis, M. (2016) "Blockchains and smart contracts for the internet of things", *IEEE Access* 4, 2292–2303.

[37] Baliga, A. (2016) "The blockchain landscape", *Persistent Syst* 3(5).

[38] Sirmon, D. G., Hitt, M. A., Ireland, R. D., and Gilbert, B. A. (2011) "Resource orchestration to create competitive advantage: breadth, depth, and life cycle effects", *J Manage* 37(5), 1390–1412.

[39] Adner, R., and Kapoor, R. (2010) "Value creation in innovation ecosystems: how the structure of technological interdependence affects firm performance in new technology generations", *Strategic Manage J* 31(3), 306–333.

[40] Demil, B., and Lecocq, X. (2010) "Business model evolution: in search of dynamic consistency", *Long Range Plann* 43(2–3), 227–246.

[41] Johnson, M., Christensen, C. M., and Kagermann, H. (2008) "Reinventing your business model", *Harvard Bus Rev* 86(12), 50–60.

[42] Osterwalder, A., and Pigneur, Y. (2010) *Business Model Generation: A Handbook for Visionaries, Game Changers, and Challengers*, Hoboken, NJ: Wiley.

[43] Chesbrough, H. (2010) "Business model innovation: opportunities and barriers", *Long Range Planning* 43(2–3), 354–363.

[44] Zott, C., and Amit, R. (2013) "The business model: a theoretically anchored robust construct for strategic analysis", *Strategic Org* 11(4), 403–411,

[45] Messerschmitt, D., and Szyperski, C. (2003) *Software Ecosystem: Understanding an Indispensable Technology and Industry*, Cambridge, MA: MIT Press.

[46] Samdanis, K., Costa-Perez, X., and Sciancalepore, V. (2016) "From network sharing to multi-tenancy: the 5G network slice broker", *IEEE Commun Mag* 54(7), 32–39.

[47] Wang, J., Conejo, A. J., Wang, C., and Yan, J. (2012) "Smart grids, renewable energy integration, and climate change mitigation - future electric energy systems", *Appl Energy* 96, 1–484.

[48] Pereira, A. C., and Romero, F. (2017) "A review of the meanings and the implications of the industry 4.0 concept", *Proc Manuf* 13, 1206–1214.

3 Amalgamation of Blockchain, IoT, and Big Data by Using Distributed Hyperledger Framework

G. S. Pradeep Ghantasala
Chitkara University

Anuradha Reddy
Malla Reddy University

M. Arvindhan
Galgotias University

CONTENTS

DOI: 10.1201/9781003081180-3

3.1   INTRODUCTION

Coloration suspenseful emergence about a new epoch, blockchain technology is a pioneering origination in distributed computing technology. It was first developed as a fragment of bitcoin's fundamental substructure in 2008 [1], and its probable request spreads distant outside cardinal coins and fiscal resources. The skill is still in its initial phases and is yet to reach general and innovativeness acceptance. As the expertise grew widespread acknowledgment in recent years, there has been an outbreak of progressions, novel use cases, and requests [2]. The range of potential claims of blockchain technology is boundless, from digital currencies to blockchain-enabled authorized indentures [3] with the most capable requests yet to be advanced. The agile development in shrinking, microchip technology, and wireless communication types of machinery has subsidized unique events in the world. This has caused a surge in the sum of appropriate automated strategies for several areas, a decrease in their manufacture prices, and a model modification from the actual creation addicted to the cardinal creation. Thus, the mode in which we cooperate and with the situation has altered, utilizing existing technology to improve a well empathetic of the country. The Internet of things (IoT) has arisen like outdated pieces of machinery beginning with wireless sensor networks toward radio frequency identification, to facilitate afford competencies toward intellect and stimulate through with connecting via Internet [4]. Currently, awn IoT maneuver would subsist a computerized device from computing to a computer apparatus growth display place. The variety of requests wherever it can be cast-off to comprehend various zones of the world. The IoT acting as a vital role in revolving existing towns into keen towns, electrical attics into keen lattices, and dynasties into keen dynasties, and this is the first foundation. Rendering to several investigate bits of intelligence, the figure of associated appliances is prophesied in the direction of spread everyplace beginning with 20–60 zillion in 2020 [5] mostly owed to the enormous numeral of machines to facilitate the IoT can residence taking consign the part.

The arrival of blockchain technical knowledge conveys the openings within overwhelming the beyond confronts of IoT and the above Figure 3.1 shows the architecture of blockchain. A blockchain is fundamentally contained dispersed record dispersal that concluded the complete dispersed scheme. Through the dispersed consent, blockchains receptacle permit an operation to ensue also be authenticated within a respectively-distrusted sparse system deprived of the involvement of the responsible third party.

3.2   BLOCKCHAIN CHALLENGES

When organizations implement novel technologies, the framework of that technology plays a vital role. In what way public pact through the physical possessions of innovative expertise is knowledgeable through their prior knowledge of consuming

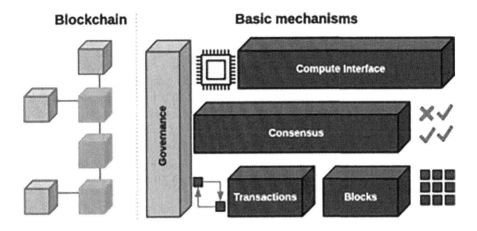

FIGURE 3.1   An outline of blockchain architecture.

or non-using alike pieces of machinery in the earlier. Meanwhile, blockchain is immobile, unique expertise, in what way organizations accept this skill, and hinge on by what means current and interrelated contests are determined. These disturb in what way organizations smear blockchain and shrewd bonds, then whether the strategy besides policymaking competencies inside can alter by the petite announcement.

Necessarily, a blockchain might be seen as a dispersed record: a consecutive restraint of "blocks" where each block encompasses a history of valid network action since the last block was added to the chain. Blockchain is a picture-aces accompaniment to IoT through enhanced interoperability, confidentiality, safety, consistency, and gullibility. Seven challenges that have to be fixed in this technology are discussed as follows:

1. Flexibility
2. Operational speediness
3. Decentralization
4. Deficiency of aptitude
5. The environment
6. Liveliness ingesting
7. Flexibility, irreversibility, quantum computing, and deficiency of standards

Flexibility: The first contest is the practical flexibility of blockchain, which is, as a minimum, intended for public blockchains, an obstacle that can edge their espousal. For illustration, the bitcoin blockchain is rising at 1.5 MB for each block every 10 minutes. It now partakes 250 GB, whereas an Ethereum complete record bulges at present receipts up and around over three terabytes of information. Lumps that need to authenticate connections be situated mostly in the direction of copying the whole bitcoin blockchain, which might stance tricky in the extended track.

Flexibility is fewer of a delinquent designed for secluded blockchains, such as Hyperledger. Meanwhile, the bulges in the system partake straight attention in dealing with connections; it can be termed as the computational methods are essential on the way to authenticate chunks is a smaller amount of problem. Uncertainty connections cannot be confirmed simultaneously or within a stretch edge. This disturbs the practical implementation of a blockchain, as fast choices are frequently essential, exclusively in today's high-velocity atmospheres.

Operational speediness: The next contest is interrelated toward the operational speediness of a blockchain. In 2018, the bitcoin blockchain was immobile accomplished of the only dispensation seven connections another, whereas the Ethereum blockchain possibly will, tentatively, process eight. Nevertheless, at the 2018 Singles' Day in China, Alibaba administered 330 transactions per second. Blockchain will yield time to spread these levels. In the intermediate, novel dispersed ledger skills are in existence technologically advanced that proposal thousands or unfluctuating millions of connections respectively subsequent.

Decentralization: The tertiary contest encompasses the level of devolution employing the bitcoin blockchain even though this does not smear to all dispersed record knowledge. The aforementioned remains significant to a high spot. The influence of bitcoin deceits cutting-edge the datum that it was premeditated in a decentralized direction, in that not one federal investor might regulate the system. Nevertheless, today's removal pools, on the other hand, switch a widely held of bitcoin's shared mess rate. Six removal pools organized switch over 78% of the removal power. This concentration of authenticating transactions is a rational reputation by which means the bitcoin etiquette was innovative, as it loots financial prudence of measure. This ensures not to be a delinquent, on condition that the taking out puddles would be reliable and consume an inducement to fix the accurate entity.

Deficiency of aptitude: From an organization project viewpoint, the fourth challenge is the deficiency of aptitude to construct dispersed requests. Taming staff to effort through blockchain receipts spell; nevertheless, the aforementioned is not yet trained by the side of several educational institutes. Only 50% of the world's topmost academics proposed blockchain courses. By way of instance, through completely novel technologies, establishments, and university circles essential to effort organized to certify the precise route is announced.

Here are previously hundreds of blockchain start-ups, all annoying in the direction of fascinating the identical inadequate aptitude, so far establishments are confronted through aptitude puddle that is escalating further deliberately than request is rising. Hence, organizations that need to hold blockchain technology prerequisite to have bottomless pouches to recompense for professional earnings or depend on staff that has only just on track erudition this unique expertise.

The environment: The fifth contest is that a decentralized environment neighboring blockchain and subordinate dispersed crops and amenities are needed. It comprises dispersed cloud storing dispersed archiving, dispersed communication, and dispersed province name servers. Utmost of these bits of knowledge are not yet wholly established, consequential in substantial hazards for anybody who needs to become intricate with blockchain and progress entirely dispersed and self-directed organizations. Nevertheless, the dispersed environment comprises of many layers, of which various segments are immobile underneath progress (Figure 3.2 indicates the blockchain environment):

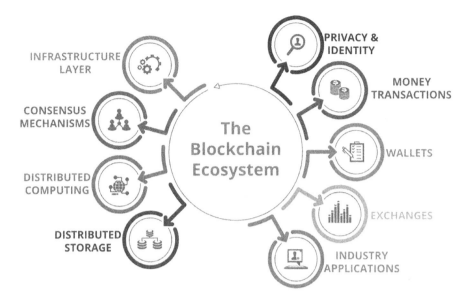

FIGURE 3.2   The blockchain ecosystem.

1. **Infrastructure layer**: Individuals request that intention to generate an organization coating on which others might advance requests. Communal blockchains consist of Ethereum, EOS, and NXT, while private blockchains consist of Ripple, Hyperledger, MultiChain, and Chain.

2. **Consensus mechanisms**: Essential to guarantee the public of the Web and to adjust which bulge would authenticate connections. Here are several consent procedures accessible, oscillating on or after testimony of employment, the evidence of venture, and several others.

3. **Distributed computing**: They are utilizing distributed ledger knowledge to dispense your calculating necessities. Fundamentally, for calculating nevertheless formerly dispersed. Examples consist of Golem and Sonm.

4. **Distributed storage**: Dispersed information stowage is individually significant as soon as you need to be undisputable that information can continually be situating retrieved, irrespective of limitations some nations have. Examples consist of Storj, IFPD, and Filecoin.

5. **Privacy and identity**: Facilities that remain absorbed on emerging a self-sovereign uniqueness confirm that information of cyberspace workers is reserve desolated and individual. Examples consist of Sovrin, uPort, Civic, and Blockstack.

6. **Money transactions**: There are four different categories of tokens: coinage, refuge, effectiveness, or quality signs. Coinage signs, connotation cryptocurrencies, are used to make monetary connections, and the furthermost eminent is, of the path, bitcoin. Others consist of Zcash, Bitcoin Cash, or Monero.

7. **Wallets**: Indeed, all those cryptocurrencies prerequisites to be set aside somewhere. Folders are the bank accounts of the crypto world. You can

have sizzling older or taciturn folders. Examples consist of MyEtherWallet, Jaxx, Exodus, or Trezor.

8. **Exchanges**: Similar to frameworks in businesses, demonstrations want to be swapped; consequently, there is a variety of federal and dispersed interactions. Federal interactions have the hazard of being scythed, which is not thinkable with a scattered discussion. Examples of regulatory interactions consist of Bitfinex, Bitstamp, Coinbase, or Kraken. Illustrations of dispersed exchanges consist of 0x, bisque, bit shares, or Ether Delta.

9. **Industry applications**: Each trade could custom DLT to expand partnership, permit origin, speed up transaction reimbursements, or permit shot.

Liveliness ingesting: The sixth contest apprehends the vigor in the getting of dispersed systems. Even though there is a variability of agreement machinery, the Proof of Work (PoW) agreement mechanism is immobile the utmost castoff. PoW needs resolving tricky riddles, which practice enormous quantities of liveliness. It is estimated that the PoW agreement machinery in the bitcoin blockchain formerly consumes 66.7 TW-hours per year, which is similar to the entire liveliness ingesting of the Czech Republic, a nation of 10.8 million public.

Fortuitously, novel blockchains capacity usage changed agreement machines, which necessitate minimal liveliness knowingly. In current years, the amount of accessible agreement algorithms has shattered. It appears that each blockchain is emerging its agreement machinery. Here is a summary of what is out there:

Vicarious testimony-of-venture: Equivalent as a testimony of venture, nevertheless the figure of indications you individual regulates who becomes to poll and those witnesses;

Leased proof-of-stake: Workers can advance their indications, which they can use to expand their safety on their server ranches;

Testimony of beyond time: Alike to a PoW algorithm, however, the variance is that this procedure emphasizes further the period of the calculation;

Beginner's Byzantine fault tolerance: There is one validated that can pack numerous transactions to generate a novel block;

Deputized Byzantine fault tolerance: This contract device practices inclined concept to authenticate lumps, among specialized workers;

Absorbed acyclic graphs: They do not have a blockchain erection, nonetheless frequently necessitate employers to authenticate two connections unknown they request to add one operation. This authentication might use a necessary PoW algorithm;

Testimony-of-activity: An amalgamation of the testimony of exertion and evidence of stake to ensure that indications obtainable as payment is on period;

Proof-of-importance: The extra you direct and collect connections on the blockchain, the additional indications you will accept;

Proof-of-capacity: Used especially for dispersed storage as it exploits the accessibility and dimensions of stowage intergalactic on a user's drive;

Proof-of-burn: Mineworkers partake to demonstrate that they blackened indications, which resources transfer them to provable unspendable reports;

Proof-of-weight: Comparable to Proof of Stake, but it cruces on several other variables, called "weights", which essentially means joining various topographies of various accord algorithms.

Flexibility, irreversibility, quantum computing, and deficiency of standards: There are also tasks interrelated to information on a blockchain. Flexibility and irreversibility are binary key characteristics of blockchains; once data or connections are affixed and acknowledged by a Web, they can no longer be altered. Nevertheless, only genuineness can be confirmed through a blockchain, not consistency and accurateness. If corrupted information is presented appropriately, it will close up on a blockchain; similarly, if a file encompasses false data but is existing accurate, it will close up on a blockchain.

Hypothetically, information on a blockchain will be there indeterminately, but the growth of quantum calculating means the cryptography used currently might not be protected in the future. Consequently, information supremacy determination only surges in reputation inside organizations accepting blockchain. Deprived implementation of shrewd conventions and, hence, unfortunate computerized executive, can later outcome in incredible glitches.

3.3 BLOCKCHAIN OPEN ISSUES

Nevertheless, in various blockchain practice offers, there is no akin federal bash that receipts concern in favor of the providing facilities or reins related datasets. As an alternative, an individual get-together in the blockchain association grasps a replica of information, somewhat that depending on a dominant solo bash to grip with preserve the primary model.

The main problem is that current hazards to companies via blockchain, whatever be elucidated more beneath, are as follows: blockchain systems straddling numerous authorities, crypto properties; information safety; data safety on the blockchain; and hazard of cyberattacks.

Jurisdictional tribulations: A bump of a distributed ledger may various extent places about the biosphere. It has repeatedly been hard for creating which dominions' laws also guidelines spread over to an agreed claim. Nearby is a hazard with the aim of connections completed through an association that might drop below every dominion in whichever a bump in the blockchain network is found, ensuing in an irresistible numeral of rules well as guidelines that valor smear to connections in a blockchain constructed organization.

In a communal blockchain system, it might exist significant which contemplate what rule valor smear toward connections and contemplate suitable hazard supervision that must apply. Nevertheless, with a missioner or secluded system, it has informal to generate the roughly outward appearance of permissible agenda and inside supremacy composition that can command the first law that will spread on to communication. In secluded systems, it ordains furthermore be advantageous to contemplate approximately a variety of settled argument determination procedure.

Crypto properties: The complications of spread over the current governing administration can be perceived evidently, while it originates from the use of crypto properties. We now distinguish an enormous choice of views since officials happening crypto properties, since absolute uncertainty and prohibitions in a few states [6], to additional vigilant stakeholder advice from others [7]. However, yet further republics have announced rules to fascinate extra crypto bustle [8]. These discrepancies of judgment and the subsequent drawbacks are able-bodied recognized in the illustration for Preliminary Coin hand-outs. The admiration of vending gestures by way of PCOs as an income of start-up endowment raise has shattered in the past few years. Statistics explain that nearly $21.9 billion has been hoisted up during approximately 940 PCOs. It concluded the retro since January headed for November 2018 unaccompanied, dwarfing the expanses elevated for blockchain ventures by outdated speculation investment throughout the equivalent period [9]. Nevertheless, agreed for discrepancy of official outlook on the precise permitted insinuations on a perfunctory trade, establishments to flop to contemplate at the onset even if their perfunctory vending might be acquiescent in the dominions in whatever they proposed to bid demonstrations could aspect an indecisive upcoming. Establishments could furthermore encompass to guarantee with the purpose of an auction of presentations that have restricted to consumers during their chosen dominions in tidy to eliminate the hazard of proffer spreading to authorities that are further seriously delimited or comprise absolute prohibitions happening PCOs.

Data safety: The problem of data safety and blockchain technology has been penetratingly discussed. Several experts and scholastic reviewers have appealed with the aim of blockchain technology, which have mismatched through confidentiality regulations such as the EU Overall information Safety Guideline [10]. As stated above, the innovative persistence of blockchain enables peer-to-peer connections deprived of the necessity of an indispensable gathering. In an authorization less community blockchain system, veto solo revelry receives accountability for the accessibility or safety of a specific blockchain network. The entire manipulators of the system might contain admittance to the information on the set of connections. These characteristics struggle among the prod of confidentiality laws, which necessitate the party monitoring individual information of a personage to protect the safety with confidentiality of that information for the sake of the separate or "information subject".

Data safety on blockchain: Blockchain technology is frequently discussed as "fiddle resistant" [11]; it is universal because each new cardinal "block" comprising evidence of connections is associated with every previous block. In direction to interfere among several of the records confined in a obstruct, a deceitful applicant would necessitate altering all sequential blocks in the fetter to evade exposure. Assumed that blockchain has a distributed ledger, nearby is no lone summit of disaster to facilitate deceitful applicants know how to prevail. As an alternative, they would need an enormous quantity of influence to predominate and modify each node simultaneously.

Hazard of cyberattacks: Regardless of the elevated stage of safety to blockchain systems available to the information documented on conservatives, there are a few essential cybersecurity hazards that persist. The particular task to distributed systems, mainly community blockchains [12], is that information contribution can be commencing whichever numeral of nodes, sensing nearby is a hazard of meddling at a piece of the knob. The advantage of via "interfere testimony" expertise is invalid, but the data stockpiled on the record are conceded to commence among. This kind of occurrence is not intended at the blockchain itself, although at outside schemes such as cryptocurrency cases. At hand is a hazard that entities valor targets the information contribution idea relevant to the propagation of enormous data.

3.4   BLOCKCHAIN DISTRIBUTED HYPERLEDGER FRAMEWORK

The Hyperledger Scheme is a combined exertion to make an enterprise-grade, non-proprietary dispersed ledger framework [13] and encryption support. It resolves for the development of blockchain technology with recognizing and apprehending a crosshatched exposed standard podium in support of dispersed ledgers, which can convert the mode professional connections are directed worldwide. Conventional as a venture of the Linus Background work an initial 2015, the Hyperledger development presently has an additional than 60 members.

3.5   IOT CHALLENGES

IoT commonly highlights the flexibility [14] of applying calculating and Web skills to diplomacies and radars that aren't measured with supercomputers, allowing them to create machine-to-machine connections with nominal or nil person involvement. As the decision on the route to a price reduction for provisioning, liveliness investments, worth additional amenities, and competency of organisation and use of apparatus and substructure survives, this understanding has cautiously persuaded.

IoT diplomacies with restricted functionality have been everywhere for a minimum of an epoch. What did you say has altered newly is the ubiquity of connectivity possibilities, cloud amenities, and analytics, whatever are countless enablers for IoT. The cloud offers a stage for presenting intellectual package, schmoozing [15], a considerable numeral of IoT diplomacies, and provisioning them utilizing a significant quantity of information. It permits keen choices to be finished deprived of humanoid involvement, and Figure 3.3 shows the cloud Internet interactions.

Nevertheless, there are still some recent encounters restraining the implementation of IoT:

Safety susceptibilities: Systematic equitation of prominent goals retains this hazard continuously in the posterior of our concentrations. The significance of disruption and renunciation of amenity could be distant further thoughtful than a negotiation of confidentiality. Altering the assortment proportion of antiseptics at an aquatic handling plant or terminating the refrigeration arrangement at an infinitesimal supremacy plant might place an entire city in immediate hazard [16].

Supervisory and permissible matters: This spreads over mostly to therapeutic diplomacies, finance, indemnification, organization paraphernalia, engineering apparatus, and in specific, pharmacological and nourishment-associated apparatus. Nowadays, this callous conforming with regulations viz. CFR 21 piece 11, HIPAA, dictate 95/46/EC and GAMP 5. This enhances the period and price desirable to carry these crops against the souk.

The stoicism of the system: This is significant on behalf of practically every zone wherever IoT could be used, like rheostat claims, safety, engineering, conveyance,

FIGURE 3.3   Internet connections interacted with the cloud.

wide-ranging structure, and therapeutic diplomacies. The usage of the mist presently levies an interruption of about 250 ms or more. This is acceptable for most claims, but not for safety or other requests that need a speedy, nearly instant reply. A prompt from a safety monitoring system established 5 seconds advanced might also be twilight [17].

Deficiency of shared planning and calibration: Constant destruction in the operation of IoT will decline the price and surge the cost to the termination workers. Presently, a part of the products revealed overhead, here are furthermore Google's Brillo with Weave, AllJoyn, Higgins [18–20], to the title but a limited. Utmost of these products mark exact divisions. Some of the sources of this destruction are safety and confidentiality qualms pushing for marketplace supremacy, trying to evade problems with opponents' intellectual possessions, and the present deficiency of vibrant management to this extent.

Scalability: This is presently not abundant in an issue. Still, it is assured to develop a problem mostly in kindred to basic patron cloud as the magnitude of policies in procedure increases. This resolve surge the dossier bandwidth desirable [21] and the period desirable intended for authenticating connections.

Restrictions of the accessible radars: Essential sensor categories, such as malaise, sunlit, gesture, comprehensive, shade, sensor, optical maser, ethnography, and X-ray, are previously relatively performant. Moreover, current developments in microelectronics, attached to developments in compacted state devices, will make the unadorned methods a smaller amount of problem in the upcoming. The contest will be in creating them additional discerning in teeming, loud, and extraordinary diverse circumstances. The request for procedures that are similar to uncertain lucidity potentials to make this a smaller amount of subject in the imminent [22].

Impenetrable and sturdy off-grid supremacy foundations: Although Ethernet, Wi-fi, 3G, and Bluetooth have been capable of resolving the utmost affinity problems by accepting the various devices' variety features, these restrictions for freestyle life span immobile persist. Furthermore, smartphones tranquil essential to be stimulating every day, and utmost devices silent prerequisite steady succession vicissitudes or fitting together to the gridiron. It could put together an alteration if supremacy sources would be disseminated [23] wirelesses to such strategies since a detachment, or proviso supremacy basis that being able to latter for at least a year can be combined hooked on the radars.

3.6   IOT OPEN ISSUES

In recent years, the IoT has disappeared from a knowledge – or deposit of technologies – that be the acerbic boundary to the circumstances currently anywhere associated domestic items or vehicles, are communal. Nevertheless, development is merely actually congregation haste today by San Francisco-based Cisco, assessing that the "Internet of Everything Cisco commentary" – its acquire on the IoT – might have several as 70 billion associated devices through 2020. Figure 3.4 shows the summary of IoT.

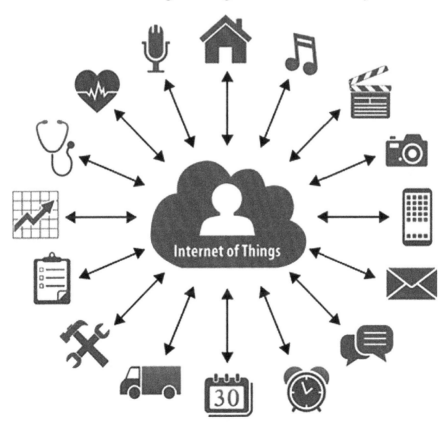

FIGURE 3.4   Summary of IoT and application areas.

As the stride of alteration quickens, and as hazard interconnections excavate, this annual year's details highpoint the increasing rinsing we are engaging in numerous of the worldwide schemes we depend on ahead. The IoT and the difficulties associated with cyberattacks [24] yield a protuberant situation in the description.

If the IoT has a problematic, or be visible to faintness, afterward, the creativities to facilitate are associated with it are similarly susceptible. In detail, despite the verity safety is unquestionably unique of the most significant issues jolt the expansion, here are a numeral of further glitches [25] to twig straight on or after this. At this time are seven main IoT glitches for visions concerning IoT.

3.6.1 CONFINED OFF INTERNET

Conferring to the global Financial Scene, the rising numeral of fractious limit outbreaks resolve to flinch assertive nationwide managements in the direction of refusal up the Internet in nationwide, or even local "enclosed grounds". There are further compressions too that will shove them to do this, counting financial tariff barriers, supervisory deviation, and the damage of rule control comparative to international online businesses.

This determination generates main glitches for the notion – and preparation of a universal IoT – foremost to the fences' manufacture to the drift of gratified and businesses. "Around valor wanted a move in the direction of a fewer hyper-globalized connected ecosphere [26], nonetheless several would not, the confrontation would be probable, as would the speedy development of prohibited work abounds. The step of technological growth would be unhurried, and its path would alter", the description recites.

3.6.2 CLOUD BOUTS

Assumed that a considerable quantity of the information that resolves track the IoT might be deposited in the cloud, it has possible to cloud benefactors will be lone of the code boards in this generosity of conflict. Although there is rising consciousness of this predicament, cybersecurity is immobile underrun during contrast to the probable age of the hazard. To find a few breeds of knowledge to the difficulty, the World Financial Scene statement quotes investigation that recommends that the transcribe of a solo cloud benefactor might source $60 billion to $130 billion of financial smashup – for feature wherever among Tornado Filthy and Tornado Katrina [27].

3.6.3 AI-BUILT SAFETY PROBLEMS

Even though the hazard extent of ransomware has grown 37 times more than that is concluded in the previous year by ransom worms, further new types of bouts are about to originate in the near future. Derek Mancy, worldwide safety tactician by Sunnyvale, Calif.-based Fortinet settles to the glitches for cloud merchants, be lone evolving.

Here though that the succeeding giant goal for ransomware is likely to be cloud facility benefactors and additional saleable amenities with an objective line of producing income rivulets. The systems connected through cloud can provide advances such as effective handing of hundreds of trades, management of different articles, safety from dangerous structures, and healthcare applications. [28].

3.6.4 BOTNET GLITCHES

Masses of novel associated customer strategies make an extensive outbreak superficial for hackers, whose spirit endures reviewing the networks among low-power, slightly quiet approach, and critical communications. Shaun Cooley, VP with CTO next to San Jose, California, founded the Cisco Internet site supposed. This initial safety test he observes is the formation for Dispersed Devastation of Service (DDoS) bouts that service clouds of poorly regulated customer strategies to bout community structure over hugely synchronized mismanagement [29] of message frequencies.

3.6.5 PARTIAL AI

AJ Abdallat has chief executive for Beyond Limits website, an association that be native on or after the laboratories of the Caltech in-depth space program. He summits out that the maximum of the present AI contributions on the marketplace encompasses significant confines. Subsequently, the machine learning plus Big Data-based AI that presently permeate be influential tackles on behalf of recognizing relations in hefty extents of information, however, don't encompass abundant on individuals during relationships for employed out the compound marvels of reason and consequence or to find adaptable issues that be capable of produce chosen results.

By way of Big Data in addition to machine learning-driven AI's advanced dispensation supremacy, they know how to consist of keen on their procedure further and added data, also and additional variables that may disturb information links. But with slight human interference, some variables may inevitably show robust association by untainted chance, with the slender, real extrapolative result.

The convenient relevance's of AI to the IoT consists of, Smart IoT to facilitate attaches and to enhance policies, information, and the IoT; AI-Enabled [30] Cybersecurity to proposals information safety encryption with improved inferential consciousness to make available file, information, and system + padlocking by smart dispersed data secured through an AI means.

3.6.6 DEFICIENCY OF SELF-ASSURANCE

Amsterdam, Netherlands-based Gemalto, has cybersecurity security that has investigated the influence of safety on IoT expansion. Uncertainty initiates that 90 out of a 100 customers absent self-assurance in the safety of the IoT policies. This originates because eclipsing two-thirds of customers and nearly 80% of institutions sustenance administration attainment tangled in background IoT safety. In detail, its new State of IoT Safety investigation description, out at the culmination of October, disclosed the following data.

Ninety-five percent of industries and 91% of customers trust in attendance must be IoT safety guidelines.

Fifty-five percent of customers have possession of the usual four IoT policies, but merely 14% trust that they are well-informed scheduled IoT maneuver safety.

Sixty-eight percent of customers were worried apropos a hacker monitoring their IoT maneuver, while 62% are worried regarding information being dripped.

"It's vibrant to equally customers along with trades have stern anxieties about IoT safety and modest self-assurance that IoT overhaul benefactors and maneuver producers will be competent to defend IoT devices and further prominently the honesty of the information formed, deposited and conveyed by these strategies", alleged Jason Hart, CTO of Data Defense at Gemalto alleged in a declaration apropos the statement. "Till there as self-assurance in IoT between industries and customers, it will not perceive conventional acceptance", alleged Hart.

3.6.7 CONSIDERATE IoT

In 2018, the actual matter was how to amplify the capability in favor of the public to recognize the variations and their allegations further visibly and to obtain real activities to benefit from the possible positive aspects. "The step of transformation can surpass the proportion of human competence to fascinate – the mug is previously chock-a-block", alleged Jeff Kavanaugh, VP, as well as Senior Partner in High Tech and Industrial on behalf of Infosys website [31].

As related tactics get shrewder and more anodyne, and opportunities to use IoT data for visions and economic assessment increase; IoT has sparked interest in its puberty. Furthermore, algorithms and information picture patterns have evolved to new usecases that can benefit from prior ones. The exponential adoption of IoT has the potential to dictate device and attainment pricing, allowing for further innovation.

3.7   IOT DISTRIBUTED HYPERLEDGER FRAMEWORK

Hyperledger is an open foundation scheme introduced by Linux Foundation. Many subprojects are derived underneath the umbrella of Hyperledger associations like Hyperledger Indy and Composer. Hyperledger Fabric is one of the schemes at first established by IBM, and the future was donated to Hyperledger. It permits us to advance secluded permission blockchain following the finest in trade values and procedures.

We are handling IoT information on the distributed Hyperledger framework. We will be gathering information from the IoT devices and securely conveying it to our node consecutively, the Hyperledger Blockchain, employing the MQTT protocol. After getting information from the device, we will process the data and enhance it to our record. We also estimate our system's presentation was numerous captivating constraints like consignment timeout, consignment size, and communication count into contemplation.

3.8   BIG DATA CHALLENGES

Information capacities are enduring to raise, and so are the potentials of what can be completed with sufficient raw information offered. Nevertheless, establishments want to be bright to recognize just what they can do with that information, and by what method they can influence to shape visions for their customers, products, and amenities. Of the 87% of corporations using Big Data [32], only 33% have been valid in data-driven visions. A 12% rise in the convenience of the information can lead to the emergence of $68 million in the disposable revenue of an establishment.

Whereas Big Data bid a lot of profits, it comes through its set of problems. This is a novel set of multifaceted knowledge, although still in the embryonic phases of growth and progression.

Some of the frequently tackled problems contain insufficient facts about the technology complicated, information secrecy, and insufficient logical competencies of officialdoms. A proportion of initiatives also aspect the challenge of a deficiency of aids for trade with Big Data technologies. Not many publics are accomplished to effort with Big Data, which turns into an even greater tricky.

3.8.1 Managing a Huge Quantity of Information

There is an enormous detonation in the information accessible. Appearance posterior a few years, and relate it with these days, and you will understand that there has been an exponential rise in the knowledge that enterprises can contact. They have information for all, precise from what a customer likes, to how they respond, to a specific trail, to the astonishing bistro that opened up in the previous city holiday.

This information surpasses the quantity of information that can be deposited and calculated, as well as regained. The contest is not so much the obtainability, but the supervision of this information. With information appealing that statistics would rise 6.8 epochs the detachment among set terrain and rhapsodize by 2020, this is undoubtedly a contest.

Laterally with growth in amorphous information, there has also been an increase in the number of information presentations – filmed, acoustic, community broadcasting, shrewd device information, etc.

Some of the modern ways advanced to accomplish this information are an amalgam of relational databases united with NoSQL databases. An instance of this is MongoDB, which is a central portion of the MEAN stack. There are also dispersed calculating systems like Hadoop to help accomplish Big Data dimensions.

Netflix is a gratified flowing display place created on Node.js. With the improved consignment of gratified and the multifaceted presentations offered on the display place, they required a stack that might knob the storing and reclamation of the information. They cast off the MEAN stack, and with a relational database archetypal, they can manage the data.

3.8.2 Safety Frights

When it originates to information, the probability for safety fissures is immense. Big Data analytics is an extremely profitable advantage for any commercial. Consequently, any corporation storing information might aspect many coercions in contradiction of the information to be robbed and used for despicable resolves. Information safety is a damage deterrence problem as well as a confidentiality one.

Inappropriately, safety fissures of such skills are not as uncommon as we may like to think, moreover. The unfortunate truth is information robbery has pointed intensely – 450% since 2011. Of more inferior quality still, information fissures are distant from a calm misfortune to rebound back from, with over half of all industries undergoing such tribulation closing down within a year.

3.8.3 INSTANTANEOUS CAN BE INTRICATE

When I say information, I'm not restraining this to the "motionless" information attainable at shared discarding. A proportion of information preserves informing every second, and establishments need to be conscious of that too. For illustration, if a merchandising firm wants to analyze buyer behavior, instantaneous information from their recent buying can assist. Few are information analysis tools accessible for similar – reliability and rapidity. They originate with ETL engines, imagining, calculation apparatuses, agendas, and other essential ideas.

It is significant for industries to preserve themselves reorganized with this information, along with the "motionless" and always obtainable information. It will aid size well visions and improve administrative competencies.

However, not all establishments can keep up with actual statistics, as they are not reorganized with the developing nature of the tackles and skills desired. Presently, there is a limited consistent tackle, however several immobile deficiencies the essential erudition.

3.8.4 LACK OF TALENTED INDIVIDUALS

There is an absolute deficiency of trained Big Data experts available at this time. It has been declared by various enterprises looking for improved use of Big Data and building further operative information investigation schemes. There is a deficiency experienced by people and specialized data scientists or data analysts existing at present day, which makes the "figure champing" problematic, and vision building leisurely.

Again, training individuals at the access level can be exclusive for a firm selling with new knowledge. Many areas a substitute employed on computerization results connecting machine learning and artificial intelligence to build visions, but this also receipts well-trained staff or the subcontracting of expert designers.

3.8.5 ITS ALL AROUND EMINENCE

Like any other quality, information can include an assortment of excellence. Some analytics capacity is out of date, imperfect, or varying. It can make it stimulating to advance visions that are necessarily applicable and related to your firm. Deprived information can lead to production creation deprived or untutored conclusions that can have thoughtful complications in the stretched track. This delinquent can be resolved when a corporation is prepared to take information extremely and indulge it to some degree, accurately valued. By triumph out to advisors who can deliver custom-made plans, eminence package, and wide-ranging assortment, you can safeguard that your information reproduces a similar level of eminence as the respite of your firm.

3.9   BIG DATA OPEN ISSUES

Concede to McKinsey [33], and the operative utilizes of Big Data profits 180 trans-mute financial prudence also escorts in a novel tendency of creative development. The benefit from treasured information outside Big Data is the simple, modest approach of existing initiatives. New-fangled contestants must appeal to staff who own serious aids in the supervision of Big Data. By binding Big Data, productions advance many rewards and enlarged sufficient competence, knowledgeable planned track, better-quality client service, novel products, and novel clients and marketplaces.

Through Big Data, workers not only aspect several striking openings but also happenstance contests [34]. Such complications lie in information apprehension, stowing, penetrating, distribution, investigation, and imagining. These contests must be overwhelmed to exploit Big Data, nevertheless, as the quantity of data exceeds our binding competencies. For more than a few eras, computer construction has been CPU-intense other than I/O-deprived [35]. This scheme disparity confines the investigation for Big Data. CPU recital duos all 18 months concede to Moore's Law [36–39], along with the act of disk drives duos at a similar proportion. Nevertheless, the revolving haste of the disks has enhanced merely to some extent above the previous era. Because an outcome of this disparity, arbitrary I/O speeds include increased ascetically, while chronological I/O speeds have improved progressively with concentration.

Contests in Big Data investigation consist of information discrepancy and incom-pleteness, scalability, appropriateness, and safety. Preceding to information analysis, information must be fit built. Nevertheless, because of the variation of datasets in Big Data, the well-organized illustration, admittance, and investigation of amorphous or partially arranged information are still exciting. Considerate the technique by which-ever information could be prepped is significant to advance information eminence and the investigation outcomes. Datasets are regularly massive at some GB or fur-ther, and they instigate from different bases. Therefore, existing practical records are incredibly vulnerable to unpredictable, partial, and piercing information.

Consequently, many information preprocessing techniques, together with data cleaning, assimilation, transformation, and lessening, must exist pragmatic to eliminate clutter and precise discrepancies to each subprocess aspects a dissimilar contest with deference to data-driven requests. Therefore, the upcoming investiga-tion must report the enduring problems correlated with discretion. These matters comprise encoding vast quantities of information, falling the calculation influence of encryption procedures, and smearing dissimilar encryption procedures to mis-cellaneous details.

Confidentiality [40] is the foremost apprehension in subcontracted information. Lately, some disagreements have exposed how roughly safety interventions employ data produced by people for their profits deprived of authorization. Continuously, strategies that refuge all user confidentiality worries must be advanced. Besides, instruction violators must be recognized, and manipulator information would not be distorted or oozed.

Cloud platforms encompass vast quantities of information. Nevertheless, clients cannot substantially evaluate the news as data are subcontracting. Thus, data integrity is endangered. The most critical contests in veracity are that formerly advanced shredding outlines are not stretched appropriate to such vast volumes of information. Veracity inspection is also problematic for a reason of the deficiency of sustenance given isolated information admittance and the scarcity of evidence concerning interior stowage.

Big Data have advanced such with the purpose of it may not be attached discretely. Big Data can be categorized with huge methods, earnings, and contests. Consequently, further investigation is required to discourse such problems with advance the well-organized demonstration, examination, in addition to storage of Big Data. To develop these, investigate, money stashes, personal resources, and state-of-the-art thoughts are the elementary necessities.

3.10   BIG DATA DISTRIBUTED HYPERLEDGER FRAMEWORK

We have to recognize a framework that can be used to run an assignment against drapery. There are three possibilities accessible right here and now: Caliper, Gauge, and Block Bench. Caliper feels like a usual applicant, as it creates from the Hyperledger development just like drapery. Despite the fact it is companionable with drape 1.2, it supers from certain boundaries: it provisions only a solicitation. It provisions only a single operation category per track and transfers connections non-uniformly concerning the period, and it is disposed to deteriorating with unused proceedings from top to bottom circle tariffs. As a value, the Device was pronged from Caliper, which reports some of these glitches. Inappropriately, it deficiencies compatibility.

As none of the existing frameworks is completely substantial and, in the meantime, a structure is just an instrument for consecutive trials, we decided to shape our benchmarking framework. It permits us to too contract tenders consistently at a precise amount from various customers in many networks and hearsays the output of active and terminated connections per second. We use our framework for all critical trials in future assessments.

3.11   INTEGRATION OF BLOCKCHAIN, IOT, AND BIG DATA

As the innovation in the technology from corner to corner, the world has grabbed the lightning velocity, techniques reminiscent of IoT, Big Data, and blockchain (Figure 3.5 represents the trio) encompass created their entity uniqueness and demonstrated to exist an advantage to a collection of international industries. In this epoch, approximately each techie individual is familiar with every one of them autonomously and also a lot of to be familiar with their united use and its remuneration, but for the sake of apprentice reader.

3.11.1   BIG DATA

As the forename put forward, Big Data have information which is in huge quantity. Information might be of different types, i.e., prepared information, shapeless information, or partially prepared information.

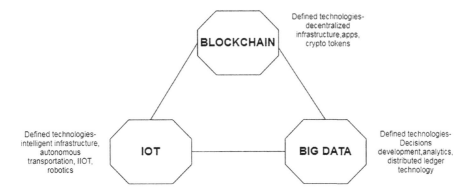

FIGURE 3.5   Blockchain + IoT + Big Data.

Big Data hold concealed pattern and procedures, which are not closed employ-ing a variety of apparatus obtainable in the marketplace. These datasets are more analyzed to offer significant business insight. Everywhere Big Data represent a tre-mendous amount of information, which is practically impracticable to development by just a single appliance despite the information is structured or unstructured. Cloud computing, on the other hand, is more than just an application that thoroughly not only supplies information and program employing a system of separate servers over the Internet but also provides services like SaaS, PaaS, and IaaS.

Several industries can professionally do well out of among Big Data analytics elu-cidations, i.e., Retail, Healthcare – Finance Services, Consumer, Telecommunication, Web and Digital Media, customer and eCommerce Service, etc., wherever information is in enormous quantity – as a result, storage space and other analytics requests to be completed by Big Data organization apparatus [41]. Besides, there are n statistics of machinery in the advertise as employing which, we be able to deal with the Big Data, i.e., Apache Spark/Storm, Ceph, Google Big Query, Hadoop.

3.11.2   INTERNET OF THINGS (IoT)

IoT looks at as the "Internet of Things" is an enormous arrangement of plans linked to the Internet. Nearly anything with an antenna on its cars, technology in manufac-tory vegetation, airliner engines, smear with oil maneuvers, wearable devices, smart-phones, tablets and further. These "things" replace information between them and exertion based on assured protocol.

IoT has to turn into the hot area of discussion in cooperation in the place of work and external. At present, at a solid phase, it carries the prospective to bang our life and employment at the equivalent point. This conception is enthralling and looks easy to attain. Still, it involves a lot of scientific and policy-related protocols that thus far are visibly implicit by the popularity of the IT production.

In easy terminology, IoT is the "Active Things" associated with the Internet. These "Active Things" might be Appliances, Vehicles, Sensors, Devices, etc. They also assem-ble and fling a few data above a few servers through the Internet, or they accept in sequence and do something leading it, or they can do together. In the forthcoming years, the obligation will be of the third category – they bring together in sequence and drive, and also they obtain some information and act in response consequently. There are thousands of relevances of such strategies in assumption which can soon turn out to be openness. Also, this kind of application will be to the enormous extent from corner-to-corner assorted industries, and the information consigned may be enormous – handled through Big Data applications. A few examples shall be in use as IoT for buildings, IoT in farming, IoT in automobiles, IoT in the healthcare segment, and so on.

3.12   BLOCKCHAIN TECHNOLOGY

Blockchain technology comes to glow among the genesis of bitcoin. Although as the moment in instance conceded, it was comprehended that the equivalent tools could be new for a lot of things, away from each other presently liability monetary connections firmly [42]. Blockchain technology includes disseminated structural design for record structure. It does not have any federal to deal with information, to a piece of certain extent information is simulated as the division of "Blocks" at various apparatus on the Internet, hence being very complicated to scythe for everyone.

Blockchains are safe by device and be an instance of a dispersed computing scheme with elevated error easiness. It facilitates reliable online connections and reports them in cryptographic form. It is a decentralized and dispersed digital ledger that indicates the links diagonally to a lot of computers so that the documentation cannot be distorted retroactively. The blockchain expertise runs on the ideology of crypto-economics. It has provided a proposal for a range of cryptocurrencies like Bitcoin, Ethereum (ETH), Dash, Ripple (XRP), and Monero (XMR). The cryptocurrencies are also called "Altcoins".

In the blockchain record, each documentation with various keys is identified as "Block". It restrains information resembling point in time squash for exacting operation and solution into preceding "Block" in sequence, which is well-simulated at a variety of location/nodes. It is set to let loose its packed possible in the upcoming years. It was the second-most often search phrase on Gartner's website with a year-long rise in exploration degree evaluating up by 450%. Tech colossal Deloitte in their full-size announcement has forecasted that blockchain may soon acquire higher than other rising technology such as cloud computing, data analytics, and IoT in business enterprise assets speculation.

Hence, it is almost unattainable to interfere with these blocks as nearby is no solo foundation of information. If anybody attempts to modify in a particular neighborhood, that alteration will create succeeding block inappropriate, so that scrupulous tamper block will be useless as of the blockchain by further nodes. In the secure phrase, as this procedure has dispersed structural design and similar in the sequence is simulated to a variety of spaces, it is practically not possible to make a change to every position/knob, so constructing this structural design a to a large extent protected individual.

3.13   CONCLUSION

IoT, blockchain, and Big Data enactment work efficaciously, which permits us to stockpile IoT information reliable and proved. Our system works fit for an attired sum of connections, but consuming just unique command restrictions its concert in rapports of the amount of operation it can hold. On Middling, our singlehanded confirmation command is intelligent to knob about 780–1600 connections reliant on several constraints like the consignment respite and consignment scope. Hyperledger platforms can be cast-off for handling IoT information, and the in-build authorizations, encryption, and passing guarantees the safety, legitimacy, and integrity of the data. Unruly technologies regularly produce considerable disagreement. Even though there are several disparagers of simulated coins, it materializes undisputable that the familiarity that withstands them is a generous industrial mutiny. Blockchain is here to stay.

Nevertheless, amending the expertise without satisfactorily warranting its maneuver circumstances wherever the price ensures not reward enhancement remains jeopardies into which individual can drop without exertion. Consequently, the assistance of spread over blockchain to the IoT must be investigated cautiously and engaged with attentiveness. It partakes providing an investigation of the first contest that blockchains with IoT have to discourse in instruction designed for them to effectively exertion simultaneously. We enclose recognized the significant facts anywhere blockchain technology can aid progress IoT claims. A valuation has moreover been providing to demonstrate the viability of blockchain bulges on IoT devices. Prevailing stands and claims have furthermore been inspected for wide-ranging training and contribution of a comprehensive summary of the collaboration between blockchain expertise and the IoT pattern.

REFERENCES

[1] Amir, Y., Coan, B. A., Kirsch, J., and Lane, J. (2011) "Prime: byzantine replication under attack", *IEEE Trans Dependable Secure Comput* 8(4), 564–577.
[2] Cachin, C., Guerraoui, R., and Rodrigues, L. (2011) *Introduction to Reliable and Secure Distributed Programming*, 2nd ed., Berlin Heidelberg: Springer-Verlag.
[3] achin, C., Schubert, S., and Vukolic, M. (2016) "Non-determinism in Byzantine fault-tolerant replication", e-print, arXiv:1603.07351 [cs.DC], http://arxiv.org/abs/1603.07351.
[4] Castro, M., and Liskov, B. (2002) "Practical Byzantine fault tolerance and proactive recovery", *ACM Trans Comput Syst* 20(4), 398–461.
[5] Clement, A., Wong, E. L., Alvisi, L., Dahlin, M., and Marchetti, M. (2009) Making Byzantine fault tolerantsystems tolerate Byzantine faults. In *Proceedings of 6th Symposium Networked Systems Design and Implementation (NSDI)*, pp. 153–168.
[6] Garay, J. A., Kiayias, A., and Leonardos, N. (2015) "The bitcoin backbone protocol: analysis and applications", In *Advances in Cryptology: Eurocrypt*, volume 9057 of Lecture Notes in ComputerScience, pp. 281–310, Berlin, Heidelberg: Springer.
[7] Liskov, B. (2010) "From view stamped replication to Byzantine fault tolerance", In (B. Charron-Bost, F. Pedone, and A. Schiper, Eds.), *Replication: Theory and Practice*, volume 5959 of Lecture Notes in Computer Science, pp. 121–149, Berlin, Heidelberg: Springer.
[8] Schneider, F. B. (1990) Implementing fault-tolerant services using the state machine approach: a tutorial. *ACM Comput Surv* 22(4), 299–319.

[9] Swanson, T. (2015) Consensus-as-a-service: a brief report on the emergence of per-missioned, distributed ledger systems. The report, available online, URL: http://www.ofnumbers.com/wp-content/uploads/2015/04/Permissioned-distributed-ledgers.pdf.

[10] Vukolic, M. (2016) The quest for scalable blockchain fabric: proof-of-work vs. BFT rep-lication, In *OpenProblems in Network Security, Proc. IFIP WG 11.4 Workshop (iNetSec 2015)*, volume 9591 of Lecture Notes in Computer Science, Springer, pp. 112–125.

[11] Yaga, D., Mell, P., Roby, N., and Scarfone, K. (2018) "Blockchain technology over-view", *Natl Inst Stand Technol* 101–106.

[12] Díaz, M., Martin C., and Rubio B. "State-of-the-art, challenges, and open issues in the integration of internet of things and cloud computing", *J Network Comput Appl* 67, 99–117.

[13] Rivera, J., and van der Meulen, R. Forecast alert: internet of things — endpoints and associated services, worldwideGartner.

[14] Che, D., Safran, M., and Peng, Z. (2013) "From big data to big data mining: challenges, issues, and opportunities", In *Database Systems for Advanced Applications*, pp. 1–15, Berlin, Germany: Springer.

[15] Hilbert, M., and Lopez, P. (2011) "The world's technological capacity to store, com-municate, and compute information", *Science* 332(6025), 60–65.

[16] intel.com/content/dam/www/public/us/en/documents/reports/data-insights-peer-research-report.pdf.

[17] http://marciaconner.com/data-on-big-data/.

[18] Luo, G., Naughton, J. F., Ellmann, C. J., and Watzke, M. (2010) "Transaction reorder-ing", *Data Knowl Eng* 69(1), 29–49. https://doi.org/10.1016/j.datak.2009.08.007.

[19] http://www.datasheetcatalog.com/info_redirect/datasheets2/19/199744_1.pdf.shtml.

[20] Rabah, K. (2017) "Challenges & opportunities of implementing blockchain security & privacy", *Mara Int J Comput Sci Inf Sec* 1(1), 1–11.

[21] Rabah, K. (2017) "Blockchain and the law: a review", *Mara Res J Law* 1(1), 22–30.

[22] Sahlberg, P., and Hasak, J. (2016) 'Big data' was supposed to fix education. It didn't. It's time for 'smalldata.' https://www.washingtonpost.com/news/answer-sheet/wp/2016/05/09/big-data-wassupposed-to-fix-education-it-didnt-its-time-for-small-data/.

[23] Smyth, D. (2016) Why blockchain? What can it do for big data? http://bigdata-madesimple.com/why-blockchain-what-can-it-do-for-big-data-2/.

[24] IoT Industry, https://builtin.com/blockchain/blockchain-iot-examples.

[25] Karimi, K., and Atkinson, G., (2013) "What the internet of things (iot) needs to beco-mea reality", *White Pap FreeScale ARM* 5, 1–16.

[26] Khan, C., Lewis, A., Rutland, E., Wan, C., Rutter, K., and Thompson, C., (2017) "A distributed-ledger consortium model for collaborative innovation", *Computer* 50(9), 29–37.

[27] Hyperledger Fabric. (2019) https://hyperledger-fabric.readthedocs.io/en/release-1.4/.

[28] Valenta, M., and Sandner, P. (2017) *Comparison of Ethereum, Hyperledger Fabric Andcorda*, Frankfurt School, Blockchain Center, Berlin, Heidelberg: Springer.

[29] Vukolic, M. (2017) Rethinking permissioned blockchains, In *Proceedings of the ACM Workshop on Blockchain, Cryptocurrencies and Contracts*, pp. 3–7, ACM.

[30] Androulaki, E., Barger, A., Bortnikov, V., Cachin, C., Christidis, K., De Caro, A., Enyeart, D., Ferris, C., Laventman, G., and Manevich, Y., et al., (2018) Hyperledger fabric: a distributed operating system for permissioned blockchains, In *Proceedings of the Thirteenth EuroSys Conference*, pp. 1–15.

[31] Sompolinsky, Y., and Zohar, A. (2013) "Accelerating bitcoin's transaction processing", Fast Money Grows on Trees, Not Chains. *IACR Cryptol EPrint Arch* 881.

[32] Stathakopoulou, C., Decker, C., and Wattenhofer, R. (2015) A Faster Bitcoin Network, Tech. rep., ETH, Zurich, Semester Thesis.

[33] 17 Blockchain Disruptive Use Cases. (2016) Available online: https://everisnext.com/2016/05/31/blockchain-disruptive-use-cases/.

[34] Nakamoto, S. (2008) Bitcoin: A Peer-to-Peer Electronic Cash System. Available online: https://bitcoin.org/bitcoin.pdf.

[35] Zheng, Z., Xie, S., Dai, H.-N., Chen, X., and Wang, H. (2017) "Blockchain challenges and opportunities: a survey", *Int J Web Grid Serv* 14(4), 352–375.

[36] Bitcoin Average Transction Confirmation Time. (2017) Available online: https://blockchain.info/es/charts/avg-confirmation-time.

[37] Finney, H. (2011) The Finney Attack (the Bitcoin Talk Forum), Available online: https://bitcointalk.org/index.php?topic=3441.msg48384.

[38] Li, X., Jiang, P., Chen, T., Luo, X., and Wen, Q. (2017) "A survey on the security of blockchain systems future gener", *Comput Syst* 1–25.

[39] Karame, G., Androulaki, E., and Capkun, S. (2012) "Two bitcoins at the price of one? Double-spending attacks on fast payments in bitcoin", *IACR Cryptol ePrint Arch* 248, 2012.

[40] Kosba, A., Miller, A., Shi, E., Wen, Z., and Papamanthou, C. (2016) Hawk: the block-chain model of cryptography and privacy-preserving smart contractsSecurity and Privacy (SP), In *2016 IEEE Symposium on, San Jose, CA, USA*, IEEE, pp. 839–858.

[41] https://www.smartdatacollective.com/big-data-iot-blockchain-benefits-of-merging-trending-trio/.

[42] https://www.houseofbots.com/news-detail/12260-1-trending-technologies-ai-big-data-iot-blockchain-ml.

4 Blockchain and Big Data for Decentralized Management of IoT-Driven Healthcare Devices

T. Saravanan
St. Martin's Engineering College

A. Ambikapathy, Ahmad Faraz, and Himanshu Singh
Galgotias College of Engineering and Technology

CONTENTS

DOI: 10.1201/9781003081180-4

4.1 INTRODUCTION

The Internet of Things (IoT) may be a growing innovation that acts as an associate degree agent that enables you to attach good, self-firing devices, and therefore, they prefer to produce economical and dynamic scenes, communication, and collaboration. In terms of security, power, and computing outcomes, these squares measure various and active business goals. Today, the Web of Things is a gift in virtually every trade. For instance, within the field of welfare and mobility, there are many things and applications that enable you to modify your daily tasks. These do not seem to be animals. Good and good boxes, good transport, frames, communicative grate, styles, stop, of course, watch the movement of the board and in different applications, these applications [1,2].

IoT devices generate an enormous quantity of knowledge, several of which may be sensitive. At large, human services and frameworks, for instance, devices that are hooked up to a patient to produce personal info, like the patient's health standards. This info is relayed to guests at the hospital at a point where associate degree is often monitored to trigger an alarm in emergencies [3,4]. As a result of protection of such devices, in addition, because of the received information, it is important to produce the IoT-part default behavior, that is because of the very fact that the IoT context of important choices supported the info found. If a malicious device has been introduced to the Web of Things, this and disruption of the correct functioning of the system will cause fateful results. Information assortment, privacy, reliableness, quantifiability, availableness, and nonnegativity are just a few of the safety parts of the Web of Things. However, verification and access management are the most lines of defense and restrictions on access to info for persons with applicable powers [5].

Exchange-validation between IoT devices associated with different factors is an integral part of protected IoT-RAM, in addition to making certain that the data is evaluated and reliable by it; otherwise, on top of being exposed to numerous threats to their security and safety, together with any unauthorised access, stealing of knowledge, data, and changes. It is expected that IoT frameworks can add a coordinated manner, with a strict minimum to outline commitments to create gadgets of various IoT frameworks that will move with one another, giving extra services. To take care that the framework has distributed security assessments [6]. Existing security parts, for example, checking details that do not seem to be appropriate for IoT frameworks, because of their uniform structure and lack of ability. For instance, the battery management unit of the essential machine can check the system, to begin with, a distinct sequence of stations that may be open for a brief amount of your time. The location generates dedicated channels that gradually merge into external foundations, and issues are typically resolved through the use of fuzzy bushings that square measure a significant organ, communication, control, storage, and, as a result, higher-up board, within the sense of a system. Within the IoT, with the constant growth of blockchain innovations, there are typically vital accounts for order confirmation and management of advantages [7].

Blockchain will increase load capability, inexorability, and therefore the ability to adapt to an internal error, creating a reliable outcome in determining a haul. The blockchain conjointly permits you to attach to good contracts that are a decent access management device for IoT devices. Additionally, blockchain, innovation, and session recording offer the inspiration for planning and managing distributed and redistributed, trusted, and secure solutions, taking into consideration the sensitivity axis of fog-driven IoT-RAM. For instance, it introduced a blockchain system that ensures the safety and strength of the association, in addition to an individual, or a car. Additionally, developers conferred in a very energy-efficient approach, handily skip accounting for IoT applications. We provide a reprieve from soft and blockchain-empowered tools for access verification and management for IoT frameworks [8], supported mist properties, process, and, as a result, requirement for the Instagram thought.

4.2 PRIVACY-PRESERVING BLOCKCHAIN

A wide variety of extra terms and conditions, yet as an outsized quantity of knowledge poses a threat to the client's security. Brobdingnagian opens exploit managing info for its customers to gather, manage, analyze, compare, and manage Brobdingnagian amounts of information. These corporations and also the services they supply believe security breaches and misuse of client info, which might cause a security risk to our customers, although they do not savvy to try to do therefore. Blockchain exchanges do not seem to be proof against these security issues. Additionally, customers have many choices for managing info, as yet safe operation of online exchanges, as well as, however, when, wherever, and thru whom, as yet what are the precise persons and elements disclosed to every exchange [9]. This downside is combined by Instagram; as a result of its knowledge that has been keeping within the record, it is long-lasting and may scale back the customer's ability to regulate their proper knowledge. This case is combined by changes in IoT, within which billions of individuals are forced to use showing wisdom, progressive and disabled, legitimate-interest protection tools that ask for to manage cyberattacks that show them one thing to Pine Tree State regarding info, and ultimately the sensitive personal knowledge of homeowners and customers. Additionally, within the IoT – namely, client security measures – this is tough to try due to intelligent merchandise and frequently act on behalf of the client while not the data and consent of the client, looking forward to more and more complicated announcements of this rule (Figure 4.1) [10].

In this case, the network intelligence and its partner organizations attempt to strengthen the protection, that is, it is marked as essential, and to discourage the concept that it is combined with completely different lines. The text describes a variety of scientific classifications of protection. Additionally, this info, which may be outlined because the ability to be told, concerning yourself, is protected on the premise of knowledge domain analysis. For instance, the analysis of the construct of private privacy includes two main areas: privacy protection and management. On the one hand, once it involves the privacy, security and protection are outlined as protective personal information from unauthorized access and conjointly as protective

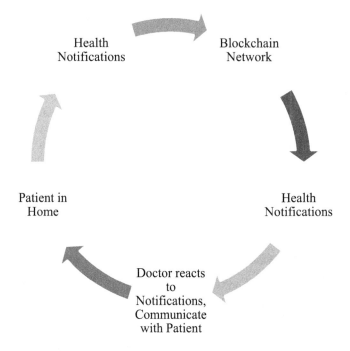

FIGURE 4.1 Monitoring of patients

personal information that is anonymous and therefore secret from the final public. In this regard, a range of tools are typically used for anonymous information to be protected, protected information and encryption info, to make sure access to the channel, and so on, to make sure dignity, privacy, disconnection, electronic security, invisibleness, and invisibleness; however with protection, this conjointly applies to native residents to properly manage and have access to info at any time guarantee the customer's trust by providing specific, low-impact certification, victimization anonymous account systems, and before revealing a restricted quantity of data to the admirer, while not revealing technical e-published numerous studies for security solutions; however, none of them are enclosed within the Instagram security tools [11]. The subsequent sections offer an outline of what blockchain is and the way it is employed in conjunction with a pet to extend client security, likewise as reflections on broad security and protection ideas like observation and privacy protection. This section also will embody the safety of storing the character of models that are delineated in Instagram, to supply customers with independence within the future.

4.3 CONCEPT OF HEALTHCARE APPLICATIONS AND BLOCKCHAIN TECHNOLOGY

In several countries, the amount of clinical patients is increasing, and it is turning into progressively tough for patients to induce and facilitate from major skilled or parent activists. With the appearance of the IoT and mobile devices, it needs up the patient's condition and kind of treatment through remote watching of the patient's condition. It additionally permits doctors to treat a lot of patients. Patients are monitored and cared for outdoor of ancient clinical settings employing a remote for patient observation (RPM) (e.g., at home). Initial of all, it permits patients to regulate their systems in an exceedingly natural means. If necessary, patients will detain bit with tending professionals. It additionally reduces clinical and money prices, while the standard of care. Usually, this can be often the primary alternative of tending professionals. World Health Organization describes it as a way of providing rate for the majority [12,13]. The rate framework, which contains very efficient oversight of the device's screen, will transmit health information to the simplest events and mobile on the network and is additionally one among the rate applications. New technologies and therefore the Web of Things (IoT) can play a very important role in the rate and progress of good town initiatives. A mobile device gathers information from a patient's medical records and transmits them to the automobile and clinical facilities to assist with health watching, designation, and treatment. This shows an oversized quantity of information, then matters, as a result of all patient information are going to be evaluated and transmitted. Transportable gadgets within the field of human services are intelligent electronic devices with multiple controllers that may be integrated into wear or worn as a dress on the body. They are invisible, simple to use, and related to advanced options like telecommuting, continuous input and early-warning systems, and integrated devices [14,15]. These devices will give vital info to social staff and repair suppliers, like stress, blood circulation, blood glucose levels, and rate, to call simply many.

This is a system of mobile gadgets, installation, programming, equipment, sensors, drives, and a network that enables you to handle info, and therefore, the market on a mobile device in check [16,17] is often one among the options of the IoT. We are not simply compatible with these devices, devices for sending social info in today's good town; however, it is a great way to simply expect mobile gadgets to be able to share a lot of information there if you connect these devices. As a result, the amount of ideas given regarding mobile, human services, devices, and blockchain innovations is increasing on the far side of what we can see or imagine. Such a structure is critical for the secure exchange {of info|of data|of knowledge} to scale back the result of such a patient – this is often information regarding several principles. Health info is deeply personal and revealing and might increase the chance of exposure. Additionally, the present information exchange framework provides an Associate in Nursing-integrated style that is essential for establishing trust and assurance [18]. Blockchain innovations are and presumably are going to be the answer to the matter of protecting the privacy and information security. Blockchain technology provides resilience to frustration and the dissemination of knowledge. A blockchain is a universal data system that stores price-supported history. As several as potential on the squares to induce to understand one another. Genesis refers to the primary field within the price chain. The header block, the number of transactions, and therefore the dealing itself are going to be contained in every one of the squares. It operates as a redistributed information storage system [19,20].

Because Instagram's security relies on the thought of an excessive amount of work, trades were deemed essential on condition that decent computing power was taken into consideration within the context of proof by the competent authorities.

Excavators were (the responsibility of the square of creation) perpetually attending to solve the puzzle (also called proof of labor (PoW) by hard the hash. Mining is that the method of adding another square to the blockchain. The hash price within the header is employed to spot every square within the provided chain. Employing a secure hashing rule designed in such a way that the hash is exclusive (SHA-256) [21,22]. SHA does not carry any text length and generates a 256-fragment coding hash. Every header contains the present state of the previous sq. within the provide chain. Blockchain is best fitted to innovation in pharmaceutical services because of the shortcoming to delete or stop working information during this space. However, integrating Instagram within the context of the IoT is not simple, and it comes with a variety of challenges, together with the necessity for top computing power for hacking, POWs, restricted flexibility, and long queues to share impressions of the system [23,24] to form it work for IoT devices, we tend to propose a replacement blockchain-based paradigm that eliminates the thought of valuation. Our model relies on the distributed nature of the system, and therefore, the other relies on further safety features.

If you share them on Instagram, they are going to be sent for complimentary below the blockchain structure and should contain further info regarding the sender and recipient. After we observe Instagram, the underlying bitcoin, and everybody features a pool address, everybody will look into what measures Associate in Nursing address in an exceeding town ought to have. As a result, the purchasers are within the system, there is nothing mysterious regarding this. We offer an easy, safe, life-saving sign ring that is excellent for secret exchange of reliable customers and lack of brightness, and thus, the client becomes a reality. For a lightweight, powerful complete, on the one hand, it ensures that the info has not been modified, because it has been protected by a well-designed label that will be broken, and therefore, the material is going to be modified with nice character, but, on the other hand, permits the nondepository financial institution to sign the message anonymously [25,26].

This is completed that is mixed with numerous Instagram (the so-called ring), and nobody is aware of; however, the dot is marked within the mail (except for the important endorser). To form it a lot of appropriate for Instagram and therefore the Web of Things, we do not place complicated tasks on the services, like admixture and powering in-chain markup. Additionally, we will use lightweight coding calculations (PLAYER number) and secret writing the secret with that you would like to code the info. Double-secret writing, which suggests that for the primary time cruciform coding and scrambling are used throughout information secret writing, the cruciform secret is used exploitation the open button. If this is often not too tough, keep in mind that we will not embrace constant info in two completely different analyses [27,28]. The PLAYER may be a form of coding, calculation ways supported easy three acts, rational enlargement, Instagram amendment, and answer OR. The PLAYER character has generated plenty of interest and tilt in each business and tutorial circle over the past few years. The Diffie–Hellman key exchange method is to possess a reliable thanks to trade crypto keys on channel discovery. Each of those ways is used along, security and protection, protection, and that, as a shopper, information is protected with lightweight techniques appropriate for little IoT devices.

4.4 CHALLENGES OF IOT AND BIG DATA

While there are many major edges of IoT within the tending trade, there are many steps that require to be taken. Lack of recognition of those issues, a network of things that healthcare suppliers must not contemplate before mistreatment. The contribution of generated data is gigantic. Having an oversized range of gadgets, all in one, human, services and offices, and thousands more, and receiving information from a far off location, and this is often all at an equivalent time, they are going to generate Brobdingnagian amounts of knowledge, since the information obtained by IoT in social, once making a list of necessities, is probably going to grow dramatically, from terabytes to petabytes. With the correct use of AI computing and processing within the cloud, which is related to the characteristic and collection of these data, the system needs loads of development time [29,30]. As a result, building a large-scale IoT in a very pharmaceutical setting needs a major quantity of your time and energy. IoT devices enable you to increase the attack. IoT and human services offer a variety of advantages for the firm, however conjointly produce a variety of sensitive security problems. Developers have the flexibility to possess, network-connected medical devices and also use and modify data [31,32]. They will conjointly go even any and attack a medical clinic, a network, infecting Web of Things devices with the foremost noted ransomware. Patients and pulse monitors, social group blood circulation, readers, brains, and scanners that individuals suppose ought to be unbroken as prisoners – the present programming structure is noncurrent.

Many clinic IT infrastructures do not meet fashionable necessities. They are going to not be taken into consideration once wrongfully connecting to IoT devices. As a result, Kabinetov remodeled his IT follow and introduced new, additional advanced programs. They are going to conjointly get to benefit virtualization (technologies like Software-defined networks (SDN) and Network Function Virtualization (NFV)) as high-speed remote and mobile systems like advanced LTE or 5G [33]. Patients and suppliers are perpetually connected to varied health service structures, through complicated social structures that affect electronic health records (EHRs) and also the use of performance enhancements like IoT and in-depth information [34,35]. This structure has conjointly been joined to the employment of the Internet and varied kinds of drug innovations that may be worn in a very constant mode of group action. However, many steps are required to be thought-about for advanced services provided by people that may kind stable, flexible, and practical police work systems. Next, we will check up on a number of the prevailing barriers to the widespread adoption of advanced medical services [36,37].

Client security, also known as security, is also in danger of being attracted by IoT devices. Unauthorized access to IoT devices will cause a significant threat to patients' health and also their personal information. Clinical and mobile phones, among alternative devices, record outline information, procedures and movements, and clinical information within the cloud. Tag, copy, mock, HF-cross, and a cloud survey indicate that the contraption has low exposure. The cloud was there, and traffic is redirected to the right injection sequence into the device through a man-in-the-middle attack. For direct association, the attack uses the assistance of alternative phenomena, like general fit-and-play, or Bluetooth Low Vitality (BLE) practicality, to find and manage

IoT gadgets [38]. Denial of Service (DoS) attacks and might have a positive impact on the structure of human services and tolerance levels. Each of the surplus and also the use of multiple devices within the system are the most barrier to achieving the task, and within the social setting, and the accumulation of assets, this is often not invariably potentially thanks to the very fact that some components that are designed into life are basic. Thanks to the dimensions and quality of code creation, as devices for vulnerabilities, characteristic potential threats to security and health, have been and still are a challenge. This drawback is growing together with the growing range of devices connected to the Internet [39,40]. Today, the quality claim is "queen-size", with untrusted Web sites and interfaces to feature the attack. Additionally, there has recently been a rise in mobile devices (including varied kinds of sensors that may be placed and embedded in testing devices). Thanks to the dearth of security on these devices; because the presence of enormous crawler Web sites that create it straightforward to spot Internet-connected devices, these moveable devices shield against a good vary of threats. Recently, several alternative systems and enhancements have appeared within the public sector, together with Wi-Fi, BLE, and Zigbee, which may be accustomed to produce a network with a good variety of clinical devices and sensors [41]. The remote and device, however, should be shielded from eavesdropping, Sybil attacks, funnel attacks, and sleep deprivation attacks. To confirm safety, security, and protection, special attention is paid to gather knowledge, personal information, family, ancestors, electronic clinical records, and genetic information that should be shielded from developers and integrated systems [42,43]. Doctors are involved regarding classification and protection. Patients ought to have an interest in sharing their medical records as a result of the sensitive nature of medical information, like malignancies or HIV check results. There is proof that the integration of connected innovations into the prevailing clinical information framework can compromise confidentiality in medical records. These security problems arise from concern laptop innovation and also the involvement of programmers. Additionally, trade analysts generally claim to be associated with health innovation, are often created as mistakes, and vulnerabilities are exploited. Once a patient intimately shared across multiple applications, the safety risks are increased. Security devices and system settings are not properly designed and will cause a risk to the security of patients and their information. The link between geographical location of the pharmacy, medications, and also the undeniable fact that this might be a profile of a personality's current health standing, raises extra issues. Another space of concern is that the use of assorted service suppliers which will have to be compelled to disclose your data to enforcement agencies [44]. If patients concern for his or her safety, this can affect the adoption and use of innovations. The systems used for information transmission are typically terribly different, and general observation from the skin world, thanks to security and protection, also as information management, is all the additional difficult [45,46].

Between the confirmation space, it is necessary that components are utilized in totally different areas to make trust, and trust in your health exchange. Shibboleth could be a single list that promotes part validation, intra- and inter-authorized managers. Shibboleth, a framework that enables cross-domain integration, has been created and with success tested at the national level. A typical framework for Instagram, that permits a consumer during an exceedingly|in a very} complicated medical scenario to demonstrate an identity supplier (IdP) and, as a result, send a support request, also as receive support from knowledgeable association (Association) [47]. Out of reading, integrated data processing, and SP-client exchange, information nature. Shibboleth concerns single sign-on, automated, spa, and eudaemonia center preparations, as an example, employing a single strategy. The Shibboleth framework is meant to guarantee} and ensure permanent installation in varied fields of automatic spa and health installation. However, thanks to the dearth of representative offices for identification functions, like service suppliers within the community, within the country, and this facilitates the work of the whole analog advanced and thriving association, not all sensible Instagram frameworks are automatic and ordered [48,49]. An absence of skills in data technology (IT) is important to finance, notably in developing countries, opposing the planned creation of boxes like shibboleth. Another vital issue that has to be taken into consideration, mutually of the foremost complicated health issues for several countries, is that the lack of interaction between countries if it involves cooperation, automation, healthcare, and ICT funds [50]. This weakness is foretold to be not solely a consequence of restricted ICT foundations or skills alone, however conjointly an absence of a method for countries to interact globally within the trade of sensitive clinical information that they are ready to enhance telemedicine and have wonderful clinical care at a distance. Tasks, as an example, are the simplest way for Liberty Alliance to attach multiple stages, also as recommendations for domains that work everybody below one roof. OpenID, name, Openliberty, the planet Wide Internet Pool, the Organization for the Advancement of Structured data Standards (OASIS), and also the Liberty Alliance project are a number of these steps and standards. The Freedom Alliance project proposal is meant to attain consistency between online principles, temperament level, and performance [51,52]. By providing open conferences and a certification program, the importance of collaboration between multiple partners ought to even be stressed. The dearth of interaction between the multistage validation space and also the instrumentation was known as potential and helpless, which could lead to information loss, and too straight-forward steering, problems, and even reverse the comparison to the inheritance framework. the freedom Alliance project any notes that in innovation policy, management, strategy, direction, also as on Liberty Street, it is necessary to verify the embrace name, confirmation between worlds. within the absence of rules, tips for consistent and progressive health care, and exchanges in varied areas during which many land areas are also situated, it is troublesome and widespread to use laptop health care and, particularly, in countries where rate management is administered, there is no it. As a result, fashionable telemedicine applications, also as patient and data sharing in varied settings, are restricted. Counter preparation and procedures for sharing data between believers have been blocked thanks to legal restrictions. within the context of finding out the Catalan automatic social protection system, contributive to the current example of a social installation system, follow these steps:

- Clinical specialists are tested in emergency clinic personnel with a shopper ID and secret phrase or x. 509 advanced last can and testament.
  - For every resolution, the hospital is distributed the protection approval nomenclature (SAML) hooked up to the Catalan health authorities.

- Seek advice from the Catalan Council of Pharmacies, pharmacies and use X. 509, automated, will, or within the client qualification (ID/secret key).
- Permit access to that whereas looking forward to the e-policy of the Catalan Council of Pharmacies then sends a SAML confirmation to the Catalan health system.
- The foundations ought to be as specified by the specialists and also the distribution of needs to be enclosed within the Catalan health service.

In the case of one certificate authority (CA) issue, X. 509, intermediate shows are often encrypted and automatic by the patient with an answer and labeling. Additionally, the provision of a SAML server platform, which can modify the exchange of authentication info to be used in automatic health managers. Rental expressions of key options and takes for tried merchandise, created potential by a SAML service engineer. Clinical or monetary info is an associate degree example of this quality. It indicates that the dealings are being performed, and additionally describes intimately however the replacement confirmation is formed and also the kind of exchange that may work, and also the specific options of the shopper, as well as the means, are used for verification. In any case, here are a number of the disadvantages of SAML:

- Classification level advanced well-being, property certificate, it depends on the dimensions of the image that it is often used.
- Targeted on separated messages, that cannot be done except once the excellent confirmation section is expounded to the matter, and can still be receptive the nonpublic key, this enables you to encourage all coding and decryption of knowledge.

The anonymity of subjects, not a nom de guerre. However, the dimensions of the SAML-based security and authentication context, to make sure that purchasers can still stay secret, should be restricted thanks to the SAML normal obstacle.

- A one-of-a-kind SAML feature is helpless against combination-based attacks during which a minimum of two villainous element targets collaborate and share information with a previous replacement, risking privacy messages.

In this case, the CA should support open up, supported the data, coding info and responsibility confirmation ought to be a part of continuing victimization the one approved proof and management related to SAML. However, the transmission vary came back to the item are often reduced to CA and guarded space. Especially, it is a fancy tending system that is supported certificate authority information during an earth science constraint, like town, town or county boundaries, nobody is ready to offer management services outside of the certificate authority.

Hypoxic-ischemic brain disease to enhance the social well-being of the population, medical aid, and safe and reliable transmission of information to an individual, certify that the information is on the road. Here it has currently enforced victimization the ways listed below: the shopper can alternately change the order to coordinate their trade, if it is a market, supported a survey [53].

Alternative patient mercantilism offers patients access to their electronic records that you just need to trace their health, in addition to confirm whether there is a proof of incorrect charge and clinical info to update them. Wherever a healthcare organization can transmit crucial info, like analysis objects, analysis results, and need of a certificate of various specialists operating with an equivalent patient, it is not supposed to figure with the market. Most analysis relies on commerce, which happens ad-lib, clinical designation, during a public organization, which needs previous info regarding the health of alternative patients. During this case, as a rule, a press release regarding however the context is a chance to access this info. Security and privacy issues are among the foremost common barriers to making sure however frameworks work. The subsequent are a number of the issues that exist HERE: the primary is that the misuse of access through approved internal affairs officers. This sometimes happens once medical organizations disclose the clinical records of patients with unauthorized persons, either as a result of a misunderstanding, or as a result of personal reasons, or in exchange for a fee. As an example, it is famed that the clinical records of celebrities and members of Parliament are frequently leaked from the Health info Management Systems (HIMSs) to the media. Secondly, unauthorized internal users UN agency will approach it, however not for records that violate the foundations. as an example, a medical clinic and workers UN agency do not directly take care of patients, or sales representatives UN agency have not been restricted from electronically aggregation info. The last cluster will be part of this cluster, names of shareholders to light the info, however, the last cluster will arrange to take revenge on their former Instagram on your GIMS protection. Third, an associate degree unauthorized offender tries to enter the context of attacks directly or by claiming to be a member of a grouping [54]. With the expansion of good-oriented businesses, this may become a significant downside and a growing threat to HIMSs. A medical clinic for safety violations will price up to $7 million, and in terms of the company's name, fines, penalties, and legal prices, among alternative things. Vital violations were committed at organizations like the anthem, CareFirst, Premera, and UCLA Health. As a result, these come gained access to a complete of 143 million records, that is 45 of the entire population of the USA. in keeping with a report printed in 2015 by the Society for Medical info Technology and Management Systems, 64% of pharmaceutical services were targeted by external hacking last year. in keeping with Blossomer, News and 90% of all humanitarian organizations are victims of fraud within the past 2 years. It is additionally compared to the financial, legal, or instructional segments, with the bulk of knowledge intrusions disbursed within the service sector of medical and clinical enterprises. within the case, the CA should support open up, supported the data, coding info and responsibility confirmation ought to be a part of continuing victimization the one approved proof and management related to SAML. However, the transmission vary came back to the item are often reduced to CA and guarded space. Especially, it is a fancy tending system that is supported certificate authority information during an earth science constraint, like town, town or county boundaries, nobody is ready to offer management services outside of the certificate authority.

Hypoxic-ischemic brain disease (CI) provides safe, reliable, and reliable electronic information exchange and information transfer services between numerous health services of the organization to enhance the well-being of birth. At the instant, it is being enforced victimization ways that are listed below: client Service, trade and coordinated mercantilism, and industry-based analysis and trade, these are samples of a number of the activities associated with client service.

Alternative patient mercantilism offers patients' access to their electronic records that just to trace their health and in addition to confirm whether there is any proof of incorrect charge and clinical info to update them. Wherever a healthcare organization can transmit crucial info, like analysis objects, analysis results, and a certificate of various specialists operating with an equivalent patient, it is not suppose to figure with the market. Most analysis relies on commerce, which happens ad-lib, clinical designation, during a public organization, which needs previous info regarding the health of alternative patients. This can be sometimes followed by a press release regarding the way to handle the power to access this event [55].

Security and security issues are among the foremost common barriers to securing Random Access Memory (RAM). The subsequent are a number of the issues that exist HERE: the primary is that the misuse of access through approved internal affairs officers. this can be sometimes done by medical and surgical services organizations that share patients' clinical records with unauthorized persons, either out of contrivance, for private reasons, or, conversely, for further advantage. As an example, it is famed that the clinical records of celebrities and members of Parliament are frequently leaked from the HIMSs to the media. Secondly, unauthorized internal users UN agency will approach it, however not for records that violate the foundations. As an example, a medical clinic and workers UN agency do not directly take care of patients, or sales representatives UN agency have not been restricted from electronically aggregation info. The last cluster will be part of this cluster, names of shareholders to light the info, however, the last cluster will arrange to take revenge on their former Instagram on your GIMS protection. Third, an associate degree unauthorized offender attempts to interrupt the system, either by directly offensive it, or by claiming to be a member of the social insurance system. Avoiding socially oriented businesses are often a significant supply of concern and are a growing threat to HIMSs. A medical clinic for safety violations will price up to $7 million, and in terms of the company's name, fines, penalties, and legal prices, among alternative things. Vital violations were committed at organizations like the anthem, CareFirst, Premera, and UCLA Health. After that, hackers gained access to a complete of 143 million records, which is 45 of the American population [56].

According to a report printed in 2015 by the Society for Medical info Technology and Management Systems, 64% of pharmaceutical services were targeted by external cyberattacks last year. In keeping with Blossomer, News and 90% of all humanitarian organizations are victims of fraud within the past 2 years. it is additionally compared to the financial, legal, or instructional segments, with the bulk of knowledge intrusions disbursed within the service sector of medical and clinical enterprises.

4.5   EXAMPLES OF IOT IN HEALTHCARE APPLICATIONS

When connected to the Web, a standard medical device will receive new information that may be valuable, give a deeper understanding of effects and trends, change remote treatment, and provide the patient bigger freedom to decide on concerning their own life and their treatment.

4.5.1 CANCER CURING

A randomized run of 357 individuals taking medication for biological process head and neck cancers was bestowed at the ASCO Annual Meeting in June 2018. The primary experiment used a Bluetooth wireless weight scale and blood circulation tribe sleeves, as a result of the chase software package, and to ensure weekly updates of patients, doctors, symptoms, and treatment responses. Compared to the management cluster of patients World Health Organization still visits their doctors, every week, patients World Health Organization received the total good management framework suffered less serious facet effects related to improper development and treatment (without supervision). Major innovations, in step with Bruce E. Johnson, chairman of ASCO (American Society of Clinical Oncology), should be reordered, once trying to find it, for each patient and their suppliers, attention, authorization, and augmented symptoms are often recognized and recognized quickly and effectively to ease the burden of treatment. The study shows that the potential advantages with good innovation are to extend patient and Dr. commitment and to watch patient health whereas inflicting lowest interference in their daily lives. As a part of the enhancements, we tend to ar currently reaching to see continue, individuals, have gotten hooked up to their home or white background for these days within the clinic, and Richard Cooper, director of digital technology at AXA surgery care, aforementioned in associate degree interview at consultancy at a conference on the longer term of care. They will resolve what are typically extremely basic problems and restore personal satisfaction. For reference, your worker is going to be rather more visible, and this is often necessary as a result of victimization technology and is liable for this.

Insulin pens, good continuous aldohexose observation (CGM), as a malady that affects one in ten adults and needs constant observation and treatment coming up with, polygenic disorder are often a wonderful platform for the event of "smart" devices. Never-ending aldohexose monitor (CGM) could be a device that permits individuals with polygenic disorder to unceasingly monitor their blood glucose levels over many days with measurements taken at regular intervals. The US Food and Drug Administration (FDA) approved the primary CGM package in 1999, and since then, several nice CGMS have appeared on the market.

Brilliant CGMs, like Eversense and race Libre, give blood glucose levels recorded in associate degree app on associate degree iPhone, mechanical man phone, or Apple Watch, permitting the user to investigate the info and determine patterns. The race LibreLink program additionally assumes remote observation of a caregiver World Health Organization can pay attention to diabetic kids and families, moreover as senior patients. Either way, the NHS is setting out to open up for next: World Polygenic Disorder Day 2018. And on Gregorian calendar month 14, the NHS declared that race Libre, like CGM, might solely be accessible as a medicine for patients with kind one polygenic disorder. It is believed that the number of diabetic patients in the European nation World Health Organization use CGM good devices is increasing from three-d to five to 20%–25% of the time. Another good device that is presently dynamical the lives of individuals littered with the polygenic disorder is that the good internal secretion pen. As a result of innovative pen or pencil solutions, like GOCap, InPen, and Esysta, you will save the number of your time, amount, and kind of internal secretion that may be indicated throughout a meal, and advice on the simplest style of internal secretion treatment at the correct time [58]. The devices will add conjunction with a smartphone app that may store data regarding the top of the day and facilitate people that stick out polygenic disorder calculate their internal secretion dose and even with GOCap), patients are going to be able to monitor all of your meals and aldohexose levels to ascertain however their diet and internal secretion intake affect their aldohexose levels.

4.5.2 Closed-Loop Insulin Delivery

The activity of ASCII text file, OpenAPS, that is Associate in Nursing Open artificial system of the duct gland and is one among the foremost fascinating areas of the Web of Things in medication. OpenAPS may be quite positive feedback of Associate in Nursing internal secretion delivery system, that is completely different from???, which, additionally to detection glucose levels within the patient's circulatory system, produces internal secretion, effectively closing the loops. Dan Lewis and her husband, Scott Leibrand, based OpenAPS in 2015, victimization the CGMP hack and also the internal secretion siphon to alter the delivery of internal secretion in her system. A circle device, and connected changes within the quantity of internal secretion siphonitis [59] through the utilization of knowledge from a CGM, a Raspberry Pi laptop.

Mechanization of internal secretion has several edges for up the standard of lifetime of folks living with the polygenic disease. By dominant an individual's blood glucose levels, and in fact by control the number of internal secretion that has to be delivered to their systems, and by doing everything potential to stay their blood glucose levels during a safe vary, you will avoid dangerous ups and downs which is additionally referred to as hyperglycemia – high aldohexose, and this is often caused because of low aldohexose. folks with the polygenic disease may additionally need to sleep from dawn to fall without concern regarding the wind of their blood glucose levels, otherwise known as nocturnal hypoglycemia.

4.5.3 Ingestible Sensors

Proteus Digital Health and eaten sensors are another example of however wakeful drug observance is used. In line with a World Health Organization report revealed in 2003, 1/2 over-the-counter medications do not seem to be taken by appointment. Their system was an endeavor to cut back this indicator, the corporate failed to build associate degree pills that rotten within the abdomen and gave you an emotional signal that appeared, utilized by the body by a device. These data are then transmitted to the phone app, which confirms that the patient has in agreement to receive their prescription, as delineate in [60].

For example, a few ataractic drug medications and medications with uncontrolled high blood pressure and kind a pair of polygenic diseases are tested to date. ABILIFY MYCITE, whereas Proteus and Otsuka Pharmaceutical Co. prescribe antipsychotics, was the foremost FDA-approved sedative, with a comprehensive follow-up, before or at the tip of 2017. Similarly, inhalers are connected, eaten sensors will facilitate patients maintain and improve the frequency of medication intake, still because the ability to drive an additional skilled oral communication with a doctor regarding treatment. Though it appears to ME that taking pills with one device is onerous, the system permits for patient choice and will refuse to supply bound sorts of data, whether or not it is doable the least bit at the instant.

4.5.4 CONNECTED CONTACT LENSES

Clinically sensible touch points are promptly mistreated by the IoT, which are to be in an exceedingly social setting. Though the conception has nice potential, science has not nonetheless worked out the way to implement it.

In 2014, Google Life Sciences, that is currently called "Truly", a subsidiary of Google's parent company, Alphabet, proclaimed that it had been developing a sensible contact center that might quantify aldohexose levels during an exceedinglylin a very} tear and supply an early warning system for diabetic patients so that they would apprehend once their glucose levels either born or accumulated to their expressed most. For this project, we have joined forces with the drug company Alcon, a corporation of Novartis ophthalmologists.

In any case, the task that appears to possess received plenty of skepticism from specialists. World Health Organization believe that activity glucose through tears was logically even, and argued that this was the top of it. Indeed, in Gregorian calendar month 2018, it had been reported that the project had come back to Associate in Nursing finish once such a protracted absence of news on the company's progress. Clinical applications of unexpected contact with the hearth foci are often, in any case, show nice activity. Indeed, it is presently operating with Alcon to possess two sensible and targeted programs geared toward resolution of the matter of long sightedness (hyperopia caused by loss of physical property within the focus of the eye), and restoration surgery. Japanese company Sensimed has developed a noninvasive lens system referred to as plectognath fish to sight changes within the eyes and actions that will cause eye disease. plectognath fish was based in 2010 and licensed metal government agency approved, that means that it had been approved to be used in promotion and sale in Europe and also the USA, and was approved available in Japan in Gregorian calendar Sep 2018.

4.5.5 THE APPLE WATCH APP THAT MONITORS DEPRESSION

Clinically, appropriate touch points are encouraged when using IoT in a social situation. Though the construct has nice potential, science has not, however, worked out a way to implement it.

In 2014, Google Life Sciences, that is currently referred to as "Truly", a subsidiary of Google's parent company, Alphabet, declared that it had been developing a wise contact center that might quantify aldohexose levels during an exceedinglylin a very} tear and supply an early warning system for diabetic patients so that they would understand once their blood glucose levels either born or augmented to their expressed most. For this project, we have joined forces with the drug company Alcon, a corporation of Novartis ophthalmologists.

In any case, the task looks to possess received loads of skepticism from consultants World Health Organization believe that activity glucose through tears was logically even, and argued that this was the tip of it. Indeed, in Nov 2018, it had been reported that the project had come back to Associate in Nursing finish once such a protracted absence of news on the company's progress. Clinical applications of sudden contact with the hearth foci are often, in any case, show nice activity. Indeed, it is presently operating with Alcon to possess two good and centered programs geared toward resolution of the matter of hypermetropia (hyperopia caused by loss of physical property within the focus of the eye), and restoration surgery. Japanese company Sensimed has developed a noninvasive contact referred to as plectognath fish to observe changes within the eyes and actions that may cause eye disease. The plectognath fish was based in 2010 and licensed cerium FDA approved, which means that it had been approved to be used in promotion and sale in Europe and also the USA, and was approved purchasable in Japan in Sep 2018.

4.5.6 Coagulation Testing

In 2016, Roche launched a Bluetooth-enabled natural action system that permits patients to watch the speed at that blood clots kind. this can be the most effective device of its kind for decoagulant patients, and self-testing looks to assist the patient keep at intervals the therapeutic vary and scale back the chance of stroke or death. With the flexibility to send results to social care suppliers and fewer visits to the middle. The device additionally permits patients to discuss their results, cue them that it is time to see them, and also the results are displayed in a very given range [61].

4.5.7 Apple's Research Kit and Parkinson's Disease

Apple added a brand-new compilation of a disorder of the API of the ASCII text file analysis Kit API in 2018, creating Apple Watch monitor Instagram disfunction symptoms. The doctor will routinely assess facet impacts with physical, vivid testing, and the patient will be requested to stay informed in order to develop a more comprehensive understanding of the phenomenon over time. The API aims to manage and program semipermanent programs.

This is the Associate in Nursing iPhone app that permits you to enter data on the graph, this daily and each hour of the accident, moreover because of the smallest changes in events. Apple ResearchKit is additionally utilized in several different studies, together with joint pain, a study conducted together with CSF and sight of study that uses sensors within the Apple Watch to trace the onset and period of seizures [62]. Apple quick regarding the app's potential to support clinical analysis and care, and since of this, in 2017 we tend to launched CareKit, that is Associate in Nursing open supply framework designed to assist engineers to develop applications for treating this illness. CareKit is most frequently used for designing applications for a particular clinical purpose, as against HealthKit, that focuses on overall well-being and prosperity, therefore you have got to traumatize this sector, with further clinical enhancements that may use iPhone and Apple Watch technologies.

4.5.8 AUTOMATED DEVICE FOR ASTHMA MONITORING AND MANAGEMENT

The Automated Device for Asthma Monitoring and Management (ADAMM) could be a transportable respiratory disease patient show that may find the first symptoms of associate asthma, permitting the user to enter and catch it before it attacks you. It will vibrate to alert the host of associate close asthma, and you will be able to conjointly send a text message to the selected leader at a similar time. Device location, the device will find and track your device usage if the patient is unable to recollect after they are accustomed it, and voice log file systems, including things like changes in thinking and habits, are two different options of the device. It conjointly includes a calculation feature that once a moment it learns that it is "normal" for someone, thus it is abundant easier for them to work out once there is one thing that must be modified. For patients with respiratory disease UN agency uses ADAMM plans, drug updates, please check the data on their phone, and this can permit you to recollect their action arrange victimization in the app and interface.

A contraption that ought to be at the agency, which ought to be free before the tip of 2017, also does not prove that gadgets take an extended time to urge back on the market, although they were created. However, consistent with Instagram's 2018 report, on the continuing health screening of competitions that use IoT gadgets, such as ADAMM, you ought to request the agency approval before long. In each trade, huge information has been modified; however, we tend to track, analyze, and value the impact of data. One of the foremost promising areas within which it will be accustomed to implement services in personnel is an improvement.

4.6 EXAMPLES OF BIG DATA IN HEALTHCARE APPLICATIONS

4.6.1 PATIENTS PREDICTIONS FOR AN IMPROVED STAFFING

For our first example of large knowledge in human services, we will cross-check a typical downside that any film director faces: what percentage individuals ought to rent at any time. If you rent an Associate in Nursing excessive variety of laborers, you risk acquisition extra labor prices. You will have terrible consumer service outcomes if there are not enough staff, which could be fatal for patients within the business.

At least during a few emergency clinics in Paris, huge knowledge is contributory within the fight against this downside. A Forbes article delves at however four emergency clinics may well be of help. The Publique-Hôpitaux Diamond State Paris is analyzing knowledge from a spread of sources to see what percentage of patients are required at every medical clinic on a daily and hourly basis.

Ten years of emergency clinic confirmations record are one in all the first informational indexes, that researchers crushed victimization time arrangement examination methods. The scientists were able to notice extensive cases of confirmation rates as a result of their study. They will then use AI to get the foremost precise calculations that foreseen future confirmation patterns. Forbes summarizes the findings: The result is a Web browser-based interface supposed to be utilized by specialists, attendants, and medical clinic organization workers – all of whom are undisciplined in scientific discipline – to forecast visit and confirmation rates for succeeding 15 days. Once Brobdingnagian volumes of guests are expected, a lot of workers is often known as in, leading to fewer hanging tight moments for patients and a bigger quality of treatment.

An EHR is the commonest application of huge knowledge in health care. Every patient has an automatic record containing sociodemographic knowledge, a clinical interview, and also the results of tests performed at the research facility. Records are divided into protected knowledge, settings, open repair suppliers, and private responsibility for gap them. Every record consists of one variable document for professionals to form changes once a definite amount of your time, and while not forms, while not the danger of information replication. EHRs may also send reminders and updates a few patients would like for a brand-new science laboratory check to trace their medications to see whether they are following their doctor's tips [63].

Although Associate in Nursing EHR may be a nice plan for several countries, it still struggles to be enforced. In keeping with this HITECH study, a North American country has created important progress, reaching 94 emergency clinics to receive EHRs; however, the EU remains insulated behind. It could be because of the strict mandate developed by the ECU Commission ought to an amendment to record European sensible that ought to become a reality by 2020. Kaiser Permanente may be a leader within a North American country Associate in Nursing function an example for the EU to follow its footsteps. We tend to sign a framework program called HealthConnect, which ought to share info across all of our offices, creating it is easier for you to use EHRs. "This integrated framework has been improved, resulting in sickness, disorder, and reaching a projected $1 billion investment in primary workplace visits and laboratory analyses", McKinsey aforementioned during a study supported solid information for men.

4.6.2 Real-Time Alerting

For our initial example of large information in human services, we will scrutinize a typical downside that any picture director faces: what number individuals ought to rent at any time. If you rent an excessive range of laborers, you risk acquisition reserve labor prices. You will have terrible shopper service outcomes if there are not enough employees, which could be fatal for patients within the business.

At least during a few emergency clinics in Paris, massive information is causative within the fight against this downside. A Forbes article delves at however four emergency clinics can be of help. The Publique-Hôpitaux Diamond State Paris is analyzing information from a range of sources to see what number of patients are required at every medical clinic on a daily and hourly basis.

Ten years of emergency clinic confirmation records are one in every of the first informational indexes, that researchers crushed exploitation time arrangement examination methods. The scientists were ready to notice the right smart cases of confirmation rates as a result of their study. They will then use AI to find the foremost precise calculations that foreseen future confirmation patterns. Forbes summarizes the findings: The result's an Internet browser-based interface meant to be employed by consultants, attendants, and medical clinic organization staff – all of whom are untrained in IP – to forecast visit and confirmation rates for consequent 15 days. Once Brobdingnagian volumes of guests are expected, additional employees are often known as in, leading to fewer hanging tight moments for patients and a larger quality of treatment.

An EHR is the most typical application of massive information in health care. Every patient has an automatic record containing sociodemographic information, a clinical interview, and therefore, the of tests are performed at the research facility. Records are divided into protected information, settings, and repair suppliers that are open, private responsibility for gap them. Every record consists of one variable document for professionals to create changes at a particular time, and while not forms, while not the danger of knowledge replication. EHRs may send reminders and updates to a couple of patients who would like to replace research laboratory check to trace their medications to see whether they are following their doctor's tips [63].

Although an EHR may be a nice plan for several countries, it still struggles to be enforced. Per this HITECH study, a North American country has created vital progress, reaching 94 emergency clinics to receive EHRs; however, the EU continues to be insulating material behind. It could be because of the strict mandate developed by the ECU Commission ought to modification to a record European smart, that ought to become a reality by 2020. Emperor Permanente may be a leader within the North American country and function as an example for the EU to follow its footsteps. We tend to sign a framework program referred to as HealthConnect that ought to share data across all of our offices, creating it is easier for you to use EHRs. "This integrated framework has been improved, resulting in unwellness, disorder, and reaching a projected $1 billion investment in primary workplace visits and laboratory analysis", McKinsey aforementioned during a study that supported solid information for men.

4.6.2.1 Enhancing Patient Engagement

It is to be expected that many shoppers and, as a result, patients are all pleased with all the good gadgets that followed their every move, their push for sleep, predisposition, and so on. Usually, this can be vital info that is often utilized in conjunction with different distinguishing info to detect hidden health hazards. As an example, an endless sleep and a speedy heartbeat could indicate a risk of developing a vascular system within the future. Patients are very fascinated by taking care of their health, and also, the encouragement of an insurance policy will encourage them to steer a healthy fashion. An alternate strategy is to reconnect and a part of the continuation of labor, in step with pronounced health patterns and transferring them to the cloud, wherever doctors will adhere to them. If any patients have an asthma attack or rate issues, they will profit, as this enables them to be additional freelance and reduce the number of unneeded visits to the doctor.

4.6.3 Prevent Opioid Abuse in the USA

The world is dealing with a sensitive topic especially in United States: this year, there are more random snippets of opioid addiction within the United States. "A Master's Study", "Bernard Marr discusses this downside true has worsened to the purpose that the North American nations have declared drug use a 'national health crisis'", and President Barack Obama has committed $1 billion throughout his presidency to handle the matter. Exploiting huge knowledge analysis in social insurance will be an answer that is searching for Heavenly, Cross and Blue protect data, scientists have started operating with fuzzy logic analysis professionals to unravel this downside. Exploiting fuzzy logic, the researchers were able to build a distinction of 742 risk factors that predict with a high degree of accuracy if an individual is in danger of addiction exploitation massive amounts of self-safety and drug safety knowledge and drug storing knowledge. "Hate could be a quality", "appealed to a specialist, and this may be the ultimate predicate, it is true", per Blue Cross and Blue protect data investigator Brendon Cosley in a very Forbes article. "It is smart that you have got reached the purpose wherever you cannot head to the doctor any longer and you have different conditions, thus if you head to a similar doctor and pay the night at a precise purpose within the unit...", "to be honest, being in reality with folks that are known as 'high risk' and preventing them from developing the unwellness, it will be risky". As a method or another, this project offers a hope for a determination of every downside that destroys several people's lives, whereas it values tons of cash.

4.6.4   USING HEALTH DATA FOR INFORMED STRATEGIC PLANNING

Huge information is used in social services, and there is a need of key indicators that help people find the simplest items of information of their interest. Units settled in a very different phase of conferences will use a clear stage in recording leads to vale care and establish those elements of them that refrain from continued treatment. The University of FL, Google Maps, and free public health information permit you to form heat maps that specialize in a spread of topics, like growth and chronic diseases. Scientists, in distinction to the present info and, as a result, numbers within the clinical, add most heat regions within the same approach. We were ready to study the method of moving them and pay additional attention to the units of measures within the most remote areas, as a result of the teachings that were learned from it.

4.6.5   PREDICTIVE ANALYTICS IN HEALTHCARE

We recently completed a biennial modern analysis of its best information; however, potential applications extend beyond more than the business. OPTUM Labs, the analysis company, has gathered data on over 30 million patients at EHRs to provide info for future analysis on the creation of tools to enhance patient care. The future goal of the company's aid services is to own a special United Nations agency that can assist you, at the correct time, to form data-driven choices and thus improve patient care. This will be particularly helpful once there are enough individuals within the United Nations agency that have a posh clinical history of a variety of undiagnosed diseases [57,64]. The new tools can predict, as an example, what individuals are in danger of developing polygenic disease and, thus, could also provide extra monitors for the superordinate board.

4.6.6 Reduce Fraud and Enhance Security

According to some studies, this business was two hundredths a lot of doubtless to own experiences and data noncontinuous. The explanation is simple: personal information that is extraordinarily valuable and profitable for misappropriated businesses. Additionally, every such breakup had emotional consequences. Since then, there are several organizations that have started mistreatment analytics to assist forestall security and safety threats by sleuthing changes within the traffic surroundings and on the Web or the other behavior that appears on. Large figures are several threats, and lots of folks believe that this may build their organizations even a lot of incapacitated than they are currently. However, with advances in security like encoding, firewall, antivirus software package, and so the necessity for security, and so the advantages, typically outweighs the danger. Similarly, it will facilitate forestall systematic and recurrent extortion and forgery cases. The review can build it easier for insurers to file claims, thus patients ought to establish the simplest proof for the effectiveness of their cases, and parental payments ought to be paid quicker. As an example, the Centers for Medicare and Medicaid Services reported that it saved quite $210.7 million in fraud cases in barely 1 year.

4.6.7 Telemedicine

Telemedicine was out there for quite 40 years agone; however solely currently, with the appearance of online conferencing, video, cell phones, external devices, and moveable devices have been doable to bring it to life. This word refers to the utilization of technology to produce medical aid at a distance from the drive. It is utilized in key discussions, in addition to remote patient watching and clinical coaching of tending professionals, as an example, a lot of direct application of specialists World Health Organization area unit ready to perform tasks, employing a mechanism, quick and continuous transmission of knowledge, in reality, while not being within the area with the patient. Doctors use telemedicine to produce personalized treatment plans, in addition on avoid hospitalization or re-screening. As mentioned on top of, mistreatment of the services of medical, informational, and analysis|research project|research is usually related to the utilization of prophetic research. This makes it doable for specialists to stop the deterioration of the patient's health, in addition on predict intensive clinical events earlier. Telemedicine helps to cut back prices and improve the character of labor, because of the work of patients far away from rescue groups. Patients do not have to attend in queues, and specialists do not waste time on spare consultations and work. Telemedicine to more increase the supply of medical aid to patients, patients will be checked and consulted anywhere and at any time.

4.7 CONCLUSION

The IoT platforms have to operate in exceedingly secure and distributed surroundings and with a strict minimum latency demand, so IoT devices is used to communicate safely and firmly with one another throughout time-dependent information transmission and exchange. During this study, a novel IoT management system ought to be offered which enables you to determine reliable communication between devices on a constant IoT platform, likewise as communication between devices of various IoT platforms. The projected system is predicated on blockchain innovation, to take advantage of cryptographical options, and is similarly temperament for each nature and time-blurring solutions to any issues. The projected element is employed in many alternative things for various situations. Additionally, the safety necessities and this model is outlined to judge and check our methodology and is its ability to fulfill these necessities. To avoid a big quantity of gain and productivity, pro re nata by the Assessment, to envision that every one of the squares, and within the long-standing time the task are to review the event of sunshine, agreements, contracts from that excavators ought to be hand-picked supported their reliableness and reliableness.

4.8   NOTES

The authors are the faculty members of Galgotias College of Engineering and Technology. The author is responsible for the views expressed in this publication and they do not necessarily represent the decisions, policy, or views of any organization.

REFERENCES

[1] Zhang, Y., and Wen, J. (2017) "The IoT electric business model: using blockchain technology for the internet of things", *Peer-to-Peer Network Appl* 10(4), 983–994. [Online]. Available: https://doi.org/10.1007/s12083-016-0456-1.

[2] Conoscenti, M., Vetro, A., and De Martin, J. C. (2016) Blockchain for the Internet of Things: a systematic literature review, In *2016 IEEE/ACS 13th International Conference of Computer Systems and Applications (AICCSA)*, pp. 1–6.

[3] Khalid, U., Asim, M., Baker, T., Hung, P. C., Tariq, M. A., and Rafferty, L. (2020) "A decentralized lightweight blockchain-based authentication mechanism for IoT systems", *Cluster Comput* 1–21.

[4] Priyanka, A., Parimala, M., Sudheer, K., Kaluri, R., Lakshmanna, K., and Reddy, M. (2017) "BIG data based on healthcare analysis using IoT devices", *MS&E* 263(4), 042059.

[5] Data aggregation scheduling in probabilistic wireless networks with cognitive radio capability, In *2016 IEEE Global Communications Conference (GLOBECOM)*, 2016, pp. 1–6.

[6] Security implications of blockchain cloud with analysis of block withholding attack, In *17th IEEE/ACM International Symposium on Cluster, Cloud and Grid Computing (CCGRID)*, 2017, p. 458.

[7] Ai, C., Han, M., Wang, J. and Yan, M. (2016) An efficient social event invitation framework based on historical data of smart devices, In *2016 IEEE International Conferences on Social Computing and Networking (SocialCom)*, IEEE, 229–236.

[8] Barnas, N. (2016) *Blockchains in National Defense: Trustworthy Systems in a Trustless World*, Alabama: Blue Horizons Fellowship, Air University, Maxwell Air Force Base.

[9] Fernandez-Caramés, T. M., and Fraga-Lamas, P. (2018) "A review on the use of blockchain for the Internet of Things", *IEEE Access* 6, 32979–33001.

[10] Ali, M. S., Vecchio, M., Pincheira, M., Dolui, K., Antonelli, F., and Rehmani, M. H. (2019) "Applications of blockchains in the Internet of Things: a comprehensive survey", *IEEE Commun Surv Tutorials* 21(2), 1676–1717. [Online]. Available: https://doi.org/10.1109/COMST.2018.2886932.

[11] Panarello, A., Tapas, N., Merlino, G., Longo, F., and Puliafito, A. (2018) "Blockchain and IoT integration: a systematic survey", *Sensors* 18(8). [Online]. Available: http://www.mdpi.com/1424-8220/18/8/2575.

[12] Liu, B., Yu, X. L., Chen, S., Xu, X., and Zhu, L. (2017) Blockchain-based data integrity service framework for IoT data, liming, In *2017 IEEE International Conference on Web Services (ICWS)*, pp. 468–475.

[13] Anirudh, M., Thileeban, S. A., and Nallathambi, D. J. (2017) Use of honeypots for mitigating DoS attacks targeted on IoT networks, In *2017 International Conference on Computer, Communication and Signal Processing (ICCCSP)*, pp. 1–4.

[14] Lee, I., and Lee, K. (2015) "The Internet of Things (IoT): applications, investments, and challenges for enterprises", *Bus Horizons* 58, 431–440.

[15] Tonyali, S., Akkaya, K., Saputro, N., Uluagac, A. S., and Nojoumian, M. (2018) "Privacy-preserving protocols for secure and reliable data aggregation in IoT-enabled smart metering systems", *Future Gener Comput Syst* 78, 547–557.

[16] Nawir, M., Amir, A., Yaakob, N., and Lynn, O. B. (2016) Internet of Things (IoT): Taxonomy of security attacks, In *2016 3rd International Conference on Electronic Design (ICED)*, pp. 321–326.

[17] Li, F., and Xiong, P. (2013) "Practical secure communication for integrating wireless sensor networks into the internet of things", *IEEE Sens J* 13, 3677–3684.

[18] Muhammad, K., Hamza, R., Ahmad, J., Lloret, J., Wang, H., and Baik, S. W. (2018) "Secure surveillance framework for IoT systems using probabilistic image encryption", *IEEE Trans Ind Inform* 14, 3679–3689.

[19] Hammi, M. T., Hammi, B., Bellot, P., and Serhrouchni, A. (2018) "Bubbles of trust: a decentralized blockchain-based authentication system for IoT", *Comput Secur* 78, 126–142.

[20] Won, J., Seo, S.-H., and Bertino, E. (2015) A secure communication protocol for drones and smart objects, In *Proceedings of the 10th ACM Symposium on Information, Computer and Communications Security*, pp. 249–260.

[21] Aloqaily, M., Al Ridhawi, I., Salameh, H. B., and Jararweh, Y. (2019) "Data and service management in densely crowded environments: challenges, opportunities, and recent developments", *IEEE Commun Mag* 57, 81–87.

[22] Baker, T., Asim, M., MacDermott, A., Iqbal, F., Kamoun, F., Shah, B., Alfandi, O., and Hammoudeh, M. (2019) "A secure fog-based platform for SCADA-based IoT critical infrastructure", *Software: Pract Exp* 50(5), 503–518.

[23] Hammi, B., Khatoun, R., Zeadally, S., Fayad, A., and Khoukhi, L. (2017) "IoT technologies for smart cities", *IET Netw* 7, 1–13.

[24] By, Gartner Says, (2016) "More than half of major new business processes and systems will incorporate some element of the Internet of Things", *Publ Janeiro*.

[25] Alkheir, A. A., Aloqaily, M., and Mouftah, H. T. (2018) "Connected and autonomous electric vehicles (CAEVS)", *IT Prof* 20, 54–61.

[26] Ahmad, A., Din, S., Paul, A., Jeon, G., Aloqaily, M., and Ahmad, M. (2019) "Real-time route planning and data dissemination for urban scenarios using the internet of things", *IEEE Wirel Commun* 26, 50–55.

[27] Baker, T., Asim, M., Tawfik, H., Aldawsari, B., and Buyya, R. (2017) "An energy-aware service composition algorithm for multiple cloud-based IoT applications", *J Netw Comput Appl* 89, 96–108.

[28] Kothmayr, T., Schmitt, C., Hu, W., Brunig, M., and Carle, G. (2012) A DTLS based end-to-end security architecture for the Internet of Things with two-way authentication, In *37th Annual IEEE Conference on Local Computer Networks-Workshops*, pp. 956–963.

[29] Jan, M. A., Nanda, P., He, X., Tan, Z., and Liu, R. P. (2014) A robust authentication scheme for observing resources in the internet of things environment, In *2014 IEEE 13th International Conference on Trust. Security and Privacy in Computing and Communications*, pp. 205–211.

[30] Lau, C. H., Alan, K.-H. Y., and Yan, F. (2018) Blockchain-based authentication in IoT networks, In *2018 IEEE Conference on Dependable and Secure Computing (DSC)*, pp. 1–8.

[31] Li, D., Peng, W., Deng, W., and Gai, F. (2018) A blockchain-based authentication and security mechanism for IoT, In *2018 27th International Conference on Computer Communication and Networks (ICCCN)*, pp. 1–6.

[32] Kshetri, N. (2017) "Blockchain's roles in strengthening cybersecurity and protecting privacy", *Telecommun Policy* 41, 1027–1038.

[33] Schwartz, D., Youngs, N., and Britto, A., et al., (2014) *The Ripple Protocol Consensus Algorithm*, vol. 5, Ripple Labs Inc White Paper.

[34] Abbas, N., Asim, M., Tariq, N., Baker, T., and Abbas, S. (2019) "A mechanism for securing IoT-enabled applications at the fog layer", *J Sens Actuator Netw* 8, 16.

[35] Rathee, G., Sharma, A., Iqbal, R., Aloqaily, M., Jaglan, N., and Kumar, R. (2019) "A blockchain framework for securing connected and autonomous vehicles", *Sensors* 14, 3165.

[36] Al Ridhawi, I., Aloqaily, M., Kotb, Y., Jararweh, Y., and Baker, T. (2019) "A profitable and energy-efficient cooperative fog solution for IoT services", *IEEE Trans Ind Inform* 16(5), 3578–3586.

[37] Tariq, N., Asim, M., Al-Obeidat, F., Zubair Farooqi, M., Baker, T., Hammoudeh, M., and Ghafir, I. (2019) "The security of big data in fog-enabled IoT applications including blockchain: a survey", *Sensors* 19, 1788.

[38] Roman, R., Zhou, J., and Lopez, J. (2013) "On the features and challenges of security and privacy in distributed internet of things", *Comput Netw* 57, 2266–2279.

[39] Mahmoud, R., Yousuf, T., Aloul, F., and Zualkernan, I. (2015) Internet of things (IoT) security: current status, challenges and prospective measures, In *2015 10th International Conference for Internet Technology and Secured Transactions (ICITST)*, pp. 336–341.

[40] Aman, M. N., Chua, K. C., and Sikdar, B. (2017) "Mutual authentication in IoT systems using physical unclonable functions", *IEEE Internet Things J* 4, 1327–1340.

[41] Baliga, A. (2017) "Understanding blockchain consensus models", *Persistent* 4, 1–14.

[42] Jayasinghe, U., Lee, G. M., MacDermott, A., and Rhee, W. S. (2019) "TrustChain: a privacy preserving blockchain with edge computing", *Wirel Commun Mobile Comput* 2019, 17.

[43] Wu, F., Li, X., Xu, L., Kumari, S., Karuppiah, M., and Shen, J. (2017) "A lightweight and privacy-preserving mutual authentication scheme for wearable devices assisted by cloud server", *Comput Electr Eng* 63, 168–181.

[44] Zhang, J., Wang, Z., Yang, Z., and Zhang, Q. (2017) Proximity based IoT device authentication, In *IEEE INFOCOM 2017-IEEE Conference on Computer Communications*, pp. 1–9.

[45] Vijayakumar, P., Chang, V., Deborah, L. J., Balusamy, B., and Shynu, P. G. (2018) "Computationally efficient privacy preserving anonymous mutual and batch authentication schemes for vehicular ad hoc networks", *Future Gener Comput Syst* 78, 943–955.

[46] Gope, P., and Sikdar, B. (2019) "Lightweight and privacy-preserving two factor authentication scheme for IoT devices", *IEEE Internet Things J* 6, 580–589.

[47] Feng, W., Qin, Y., Zhao, S., and Feng, D. (2018) AAoT: lightweight attestation and authentication of low-resource things in IoT and CPS", *Comput Netw* 134, 167–182.

[48] Esfahani, A., Mantas, G., Matischek, R., Saghezchi, F.B., Rodriguez, J., Bicaku, A., Maksuti, S., Tauber, M., Schmittner, C., and Bastos, J. (2017) "A lightweight authentication mechanism for M2M communications in industrial IoT environment", *IEEE Internet Things J* 6, 288–296.

[49] Gong, B., Zhang, Y., and Wang, Y. (2018) "A remote attestation mechanism for the sensing layer nodes of the Internet of Things", *Future Gen Comput Syst* 78, 867–886.

[50] Roychoudhury, P., Roychoudhury, B., and Saikia, D. K. (2018) "Provably secure group authentication and key agreement for machine type communication using Chebyshev's polynomial", *Comput Commun* 127, 146–157.

[51] Sultan, A., Mushtaq, M. A., and Abubakar, M. (2019) IOT security issues via block-chain: a review paper, In *Proceedings of the 2019 International Conference on Blockchain Technology*, pp. 60–65.

[52] Lee, K. C., and Lee, H.-H. (2004) "Network-based fire-detection system via controller area network for smart home automation", *IEEE Trans Consum Electron* 50, 1093–1100.

[53] Al-Turjman, F., and Altrjman, C. (2019) "IoT smart homes and security issues: an overview", *Sec IoT-Enabled Spaces* pp. 111–137.

[54] Hassija, V., Chamola, V., Saxena, V., Jain, D., Goyal, P., and Sikdar, B. (2019) "A survey on IoT security: application areas, security threats, and solution architectures", *IEEE Access* 7, 82721–82743.

[55] Wu, M., Wang, K., Cai, X., Guo, S., Guo, M., and Rong, C. (2019) "A comprehensive survey of blockchain: from theory to IoT applications and beyond", *IEEE Internet Things J* 6, 8114–8154.

[56] Lohachab, A., et al., (2019) "ECC based inter-device authentication and authorization scheme using MQTT for IoT networks", *J Inf Secur Appl* 46, 1–12.

[57] Zhang, X., Yang, L. T., Liu, C., and Chen, J. (2013) "A scalable two-phase top-down specialization approach for data anonymization using mapreduce on. cloud", *IEEE Trans Parallel Distrib Syst* 25, 363–373.

[58] Raptis, T., Passarella, A., and Conti, M. (2018) "Performance analysis of latency-aware data management in industrial IoT networks", *Sensors* 18, 2611.

[59] Goyal, T. K., and Sahula, V. (2016) Lightweight security algorithm for low power IoT devices, In *2016 International Conference on Advances in Computing, Communications and Informatics (ICACCI)*, pp. 1725–1729.

[60] Dexin, X. U., Zhenfan, T. A. N., and Yanbin, G. A. O. (2006) "Developing application and realizing multiplatform based on Qt framework", *J Northeast Agric Univ* 3, 018.

[61] Matt Broadstone, QJSON RPC, Bitbucket. https://bitbucket.org/devonit/qjsonrpc/src/master/, accessed September 8, 2019.

[62] Fotiou, N., and Polyzos, G. C. (2016) Decentralized name-based security for content distribution using blockchains, In *2016 IEEE Conference on Computer Communications Workshops (INFOCOM WKSHPS)*, pp. 415–420.

[63] Barua, M., Liang, X., Lu, R., and Shen, X. (2011) "ESPAC: enabling security and patient-centric access control for eHealth in cloud computing", *Int J Secur Netw* 6, 67–76.

[64] Reddy, A. G., Suresh, D., Phaneendra, K., Shin, J. S., and Odelu, V. (2018) "Provably secure pseudo-identity based device authentication for smart cities environment", *Sustain Cities Soc* 41, 878–885.

5 Visualizing Process Data with Reliability Data in Era of Big Data and I(4.0)

Nirbhay Mathur and Vijanth S. Asirvadam
Universiti Teknologi PETRONAS

B. Balamurugan
Galgotias University

CONTENTS

5.1 INTRODUCTION

In this developing era of engineering, Industry 4.0 (I4.0) is in fashion. It is mandatory to understand the meaning of "Data". However, only understanding data is not sufficient; it is also important to understand the related terms. In an industry where engineering systems consist of process, equipment, units, and their designs, all have to perform under controlled/monitored circumstances. To perform error-free operations, it's very essential to have data, either history or current data. This will give essence to get a new engineering field name as "Big Data".

DOI: 10.1201/9781003081180-5

Recently, all conducted research (i.e., from few years) is a great example to show the effectiveness of data. Data is collected when a process is being operated or maybe any experiment is performed. With recent improvisation in the community of the internet, it becomes easier to collect data. Internet-based devices such as sensors, wireless modules, smart devices, e-commerce, and many other devices have given a new dimension to visualize the data and store them [1]. The exponential growth of collecting and storing data in all sectors, including engineering, finance, medical, business, commerce, and many more, has made the use of Big Data. Here, some examples can be related such as medical imaging, bioinformatics, statistical prediction, and genomes [2,3]. Gaining momentum in data utilization has given more speed to accumulate the acceleration of data speed; as per IBM, 1.6 zeta bytes (1021 bytes) are available for digital data collecting and resourcing [4].

The latest technologies not only provide high speed but also are very proficient in providing high-speed computational resources (high-profile computing systems, cloud computing, parallel computing, etc.); all these are required to perform a large amount of data using the latest and more reliable visualization tools.

There are many definitions available that can define "Big Data". A meta-definition was given by Jacobs, which states "data whose size forces us to look beyond the tried-and-true methods that are prevalent at that time" [5]. A generic definition of Big Data was defined as "data that's too big, too fast, or too hard for existing tools to process" [6]. These two definitions were concluded as abstracts which left academics and business society to urge on the more specified definition to relate with processes. Later, a new proposed definition by Wu et.al. defines HACE Theorem, which states that "large-volume, heterogeneous, autonomous sources with distributed and decentralizes control and seeks to explore the complex and evolving relationship among data" [7]. International Data Cooperation (IDC) highlighted "4 V" characteristics of Big Data by defining "a new generation of technologies and architectures, designed to eco-nomically extracts value from very large volumes of a wide variety of data, by enabling high-velocity capture, discovery, and/or analysis" [8].

Process Monitoring, which is considered as the most core function in any industrial process, should follow the combination of triplet data/technology/analytics to develop and perform visualization for process data as shown in Figure 5.1 [9]. Most of the companies/industries' activity focuses on Industrial Process Monitoring (IPM), as it allows them to achieve better performance, quality, profitability, and safety [11–13].

This research paper is structured into six sections as follows: The basic introduction about "Big Data" and its application is mentioned to draw attention to process data and reliability data. Section 2 shows the role of Big Data and process data. Section 3 mentions reliability and reliability data. In Section 4, some of the case studies are mentioned based on process data and reliability data. Section 5 discusses the results, which are visualized based on process data and reliability data, and the Big Data. Last but not least, Section 6 concludes this research article, in which the main focus was to visualize and create a relationship between process data, reliability data, and Big Data for the era of Industry (I.4).

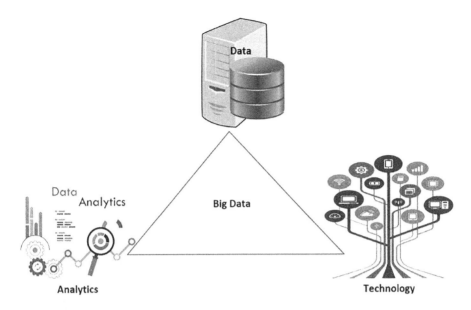

FIGURE 5.1   Big Data triplet combination for process monitoring. (Redrawn from [10].)

5.2   ROLE OF BIG DATA

Big Data is a "BIG" focus on the present era of information technology in the industry. As per McKinsey Global Institute (MGI) and McKinsey Business Technology Office, visualizing and analyzing the big amount of data will become an important key for productivity, research, innovation, and competitiveness [14].

The driver of Big Data can be classified in the following categories [15]:

• Technical
• Performance
• Productivity
• Cost

The perspective of the technology is here to have quality and the large data volume to represent the main challenges of data mining and data processing in information technology which have mostly unstructured data, especially in IoT [16]. The classical way to store data was based on Relational Database Management Systems (RDBMS); these systems were designed to deal with structured systems only since this system was not able to store or process unstructured systems. RDBM systems were designed to work with single systems, in which they were considered very adequate, but when later Big Data took place where many systems were used to work together, these systems fail to store Big Data [17].

Massive growth in recent technology has given a platform to generate data from all kinds of devices such as sensors, wireless sensors, cameras, mobile, software logs, and many others. In order to characterize the Big Data, three V's are used to do so [18]:

Volume: All sources of sending and receiving data have ever-growing data of all types. It is required to understand the cycle of volume to store data, for example, accessing data of solar power grid readings for an annual meter reading to predict for better consumption.

Velocity: Most industries prefer to use time-series data, in that conjunction is required to process data with time sensitivity since data streaming is also very famous for its fast response.

Variety: Big Data is designed to access and visualize a variety of data types such as audio, video, text, log le, and many other formats.

As per National Research Council, it is important to establish trust in Big Data technology, and results will give a conclusion for the present and future challenges, which will also make statistical learning and prediction frameworks bigger in a future scenario [19]. The role of Big Data is not limited by some applications. This technology has been used by some of the large information technology companions like Google, Microsoft, TCS, IBM, GE, P&G, and many more. A $1 billion was announced by GE to build their software, which should be expertized by using Big Data analytics [20].

The Big Data technology usage list goes from search engines (Google, Microsoft), social networking sites such as (Facebook, Twitter, LinkedIn) to the healthcare industry, mobile analytics, marketing, engineering companies, data science consultant (such as SAS, SPSS), utilities, and government. Big Data technology is equally taking part in process system engineering to integrate operation, design, and data collection to produce improved customer satisfaction by giving good efficiency and quality with all required specifications.

5.3   RELIABILITY AND RELIABILITY DATA

Big Data have features not only to have a large amount of data storage but also it can handle complicated structured and non-structured data type. Big Data can be linked with reliability and reliability data as in the modern era all plant data is collected in the digital form and it is required to have a reliability assessment.

Reliability can be explained or understood by considering the probability of the device/instrument or process, which is functioning without any failure, under all conditions for a given time of interval [21]. Reliability plays a very important and decision-making part in any process plant. It is very clear and related if higher plant reliability reduces the cost of equipment or maintenance of the plant.

Reliability plays an important role in understanding the behavior of components. It is also important to understand the hierarchic for the reliability of one instrument and the other connected to it. The system is based on the type of connections that can be broadly classified into three different categories:

1. Parallel system,
2. Series system, and
3. Mixed system.

Reliability calculation differs from one system to another system based on their component connectivity formation. An additive law is used for parallel-connected systems; in series systems, multiplicative law is used to do reliability, and for the mixed system, both additive and multiplicative laws are used. A belief reliability of a system is considered for complicated systems where hundreds of loops and multiple components are processing at the same time, such as control plants, gas power plants, machinery in industrial processing, and so on. The reliability can be measured by considering the efficiency of any system based on the quality of the units in the system. Hence, there is a direct correlation between quality and reliability.

The quality of any instrument or device depends on the following major factors:

- Material – a type of material used for fabricating the device.
- Design – architecture of the device and assembly plays an important role in ensuring quality.
- Machine
- Men power (workers)
- Management

The combination of all the above points provides the best quality, which ensures good reliability.

Reliability can be defined by $R(t)$, $t > 0$, or can be expressed explicitly as

$$R(t) = 1 - F(t) \tag{5.1}$$

where $F(t)$ is the time-depended behavior for defects or failures.

5.3.1 FAILURE RATE

The failure rate can be defined as the number of times an item fails in a given specific period to anticipate its performance. It is denoted by (t). The reliability of any equipment or component depends on the frequency of failures. This frequency of failure can be expressed as mean time between failure ($MTBF$). $MTBF$ can be represented by the following equation:

$$MTBF = \frac{T}{n} \tag{5.2}$$

where T is the total time of process and n is the number of failures that occurred during process time. Failure rate is considered as reciprocal of $MTBF$, which can be represented as follows:

$$\lambda = \frac{1}{MTBF} \tag{5.3}$$

wherein Equation 5.3, λ is the failure rate (denoted failures per 10^n hours) [22].

The failure rate can be more divided into the following categories:

- Mean time to failure (*MTTF*)
- Mean time between failure (*MTBF*)
- Mean time to repair (*MTTR*)
- Mean down time (*MDT*)

5.3.2 MEAN TIME TO FAILURE (*MTTF*)

MTTF is considered the important measure for reliability for non-repairable systems. It can be defined as the average time until the first failure equipment occurs. Mathematically, it can be expressed as follows:

$$MTTF = \frac{1}{\lambda \text{ failure } / 10^e \text{ hours}} \tag{5.4}$$

In the case of repairable systems, *MTTF* is anticipated from the occurrence of the first or next breakdown.

5.3.3 MEAN TIME TO REPAIR (*MTTR*)

MTTR can be explained as the total time spends in preventative maintenance repairs divided by the total repair. This is only implemented when there is a repairable system.

There are four major failure frequencies used for the reliability analysis as follows:

- Failure density *f(t)* – This means the first failure which is likely to occur in component or system at time *t*.
- Failure rate *r(t)* – It is a probability per unit time for the component or system which experiences failure at time *t*.
- Conditional failure intensity or hazard function *(t)* – It is a probability per unit time that can occur when a system or any component is under operation condition or repaired to perform as new with operating time *t*.

5.3.4 MEAN DOWNTIME (*MDT*)

MDT is a state when the system has not been used and includes time for all kinds of maintenance, repair, and correction in network and loop. *MDT* also includes the time for self-imposed downtime or administrative delays. Although *MDT* and *MTTR* are both confusing terms, both play a very significant role in the process industry.

If either one is used (i.e., *MDT* or *MTTR*), it is required to reflect the total time for which equipment is unavailable for service, whereas computed availability will be not correct.

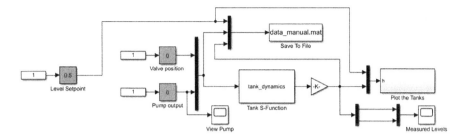

FIGURE 5.2   Simulink design for the two-tank system in MATLAB.

5.4   CASE STUDY

5.4.1   Two-Tank Process Plant

A two-tank system is designed in a simulation environment to understand the function of components. MATLAB software with Simulink toolbox was used in performing the operation and generating data.

In the above-designed (Figure 5.2) system, the rate of flow of water (or any other chemical) is used to flow from one tank to another with the help of a control valve, which plays a very important role in all process plants [24]. Control valves are the most common actuators, widely used in process regulation and process operations [24]. As per the study, more than 30% of nonlinearity problems arise due to control loops [25].

The working of the two-tank system is shown in Figure 5.3. In the two-tank system, the liquid is flowing from one tank to another tank with a controlled rate of flow

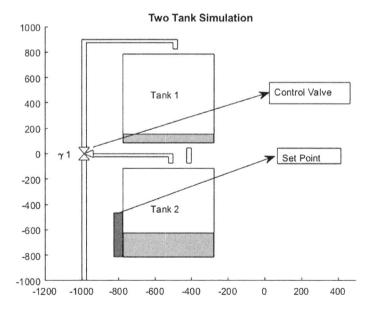

FIGURE 5.3   Two-tank system process operation designed in MATLAB. (Redrawn [23].)

with the help of a control valve. Problem arises when this system is exposed to few more systems, either by creating the series connection or the parallel connection. Also, there are many types of control valves available to be used in a big process plant to give the most promising output. Pneumatic control valves are widely used in process plants to attain the best results with lower cost and maintenance [26].

Once the system is designed and ready to generate data, it's time to understand the nature and behavior of output data.

5.4.2 Process Data Collection

As mentioned in the above section and explained about designing and performing the simulation, the data is collected as shown in Figure 5.4. The values of the rate of flow were collected and stored to visualize the trend of data. It was noticed at some places rate of flow, which either falls or has a random spike. It was concluded that these variations are considered failures in operation, resulting in the failure of data or failure in performance. The type of failure that can be developed due to the operation of control valves was considered.

5.4.3 Failure Types

To understand the failure that occurs in the control valve, it is required to understand the working of the control valve. The control valve consists of four main parts,

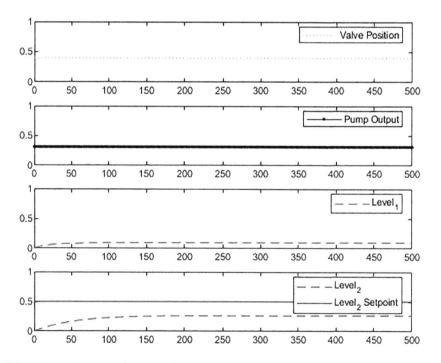

FIGURE 5.4 Output of the two-tank system.

namely, the actuator, the bonnet, the valve seat, and a valve plug. All these parts are assembled to respond to control the valve operation. Figure 5.5 shows the assembly of parts and the position of control valve works.

As mentioned above about assembly parts for the control valve, the valve plug is the most versatile part, which stays inactive position for all operations. The movement of the valve plug takes place randomly sometimes or directed steps since it creates some kinds of failure. These failures can be categorized into two types:

- Valve stiction
- Valve wear-out

Both failures occur respectively and based on the operation type. In this research, valve stiction is considered as the failure of occurrence in performance of the plant. Valve stiction is also known as static friction, which is considered a major problem in industrial processing.

To overcome this problem, valve stiction research can be divided into three parts to resolve this issue: valve stiction detection and fault diagnosis, valve stiction modeling and quantification, and valve stiction compensation.

Among these valves, stiction modeling is used as a very important method to analyze the valve stiction. It's very challenging to model a valve stiction simulation. To

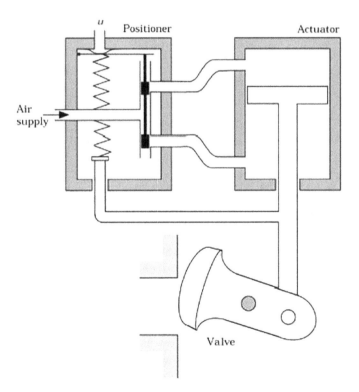

FIGURE 5.5 Working diagram of control valve.

develop a model, many factors and parameters have to be kept in mind so that it can match industry processing. At the time of the building of the valve model, it is very likely to get confused with some other control valve nonlinearities such as deadband, backlash, and hysteresis. Some researchers have made them very clear and mentioned the differences among nonlinearities that can be found at the time of valve stiction [27]. American National Standards Institution (ANSI) has defined definitions for valve stiction terms that can be referred to understand nonlinearities in deeper [28,29].

Some process control terminology/definitions are as follows as shown in Figure 5.6:

- **Backlash**: This name is given to a form of a dead band, which can result in the temporary discontinuity between the input and the output of a device. This occurs when a device changes its direction. Looseness or slack of mechanical connection is an example of the backlash.
- **Deadband**: The range through which an input signal can be varied, upon the reversal of direction, without initiating an observable change in the output signal. When the term "Deadband" is used, both input and output variables must be identified. Deadband is typically expressed as a percent of the input span.
- **Dead zone**: It is a predetermined range of input through which the output remains unchanged, irrespective of the direction of change of the input signal.
- **Hysteresis**: It is a property of an element that evidences the dependence of the value of the output, for a given excursion of the input, upon the history of prior excursions and the direction of the current traverse.

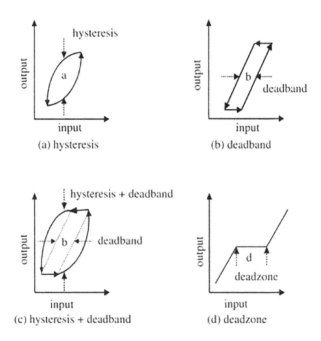

FIGURE 5.6   Control valve nonlinearities. (Redrawn from [27].)

5.4.4   RELIABILITY PREDICTION

After understanding the architecture of the system/process plant, i.e., are they connected in parallel or series and the next step is to know the reliability modeling? Once data is collected, it is required to analyze the reliability prediction. Many models are available for reliability analysis, such as exponential distribution, Weibull distribution, normal distribution, lognormal distribution, and many more.

In the late 1980s, exponential distribution was assumed to be a common method for reliability since it was having the simplest mathematical form with statistical properties. Exponential lifetime distribution is also known as the no-memory property. Later, it was found that assumptions of exponential distribution must always be taken into consideration to have a more accurate prediction for failure mechanism. Hence, different models should be utilized based on the problem, data, and variable.

From the above-listed statistical models, the Weibull distribution, which is named after the Swedish Professor Waloddi Weibull, is the mostly used distribution for lifetime analysis [30]. Weibull distribution is mostly used not only because of its flexibility in analyzing diverse types of aging phenomena, but also because of its simple and specific formation in comparison with other distribution.

So, the cumulative density function (CDF) of the standard two-parameter Weibull model is given by

$$F(t) = 1 - \exp\left[-\left(\frac{t}{\alpha}\right)^{\beta}\right] \qquad (5.5)$$

for any $t \geq 0$, $\alpha > 0$, and $\beta > 0$. Here, α and β are named as scale and shape parameters, respectively.

Now, considering Equation 5.5, to develop the failure rate function of Weibull distribution as follows:

$$h(t) = \frac{f(t)}{R(t)} = \left(\frac{\beta}{\alpha}\right)\left(\frac{t}{\alpha}\right)^{\beta} - 1 \qquad (5.6)$$

Though, the best part of using Weibull distribution is that it can have different monotonic types of hazard rate shapes so that it can be applied to different kinds of products.

As shown in Figure 5.7, a plot is sketched between the reliability parameters based on time, which shows the working degradation. Here, based on (η) values, the plot is visualized based on time intervals, in which a sudden drop has been notified when the operation comes to its end of life. This means when the process was in its initial stage, it was performing and giving the required output. In the plot, three different parameters were considered to understand the behavior of reliability failure. Three different values for reliability were estimated by using Weibull distribution and ML algorithm to predict lower value, estimated value, and upper value of estimation. These values will give a close visualization to understand the trend of failure or reliability drop.

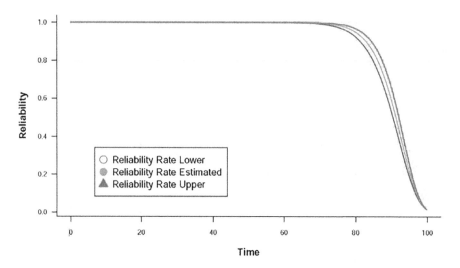

FIGURE 5.7 Reliability plot vs time, which visualizes the (η) values for operation of the control valve.

The lower estimated value will give the starting range where the failure can be just initiated in the process. It could be the first alarming stage for system engineers to check the overhaul performance. Another value is the estimated value, which shows the failure occurrence. The system should be maintained until this value or time comes into operation, although failure can be notified in the process and could stop the operation. The last parameter gives the upper limit for range, which denotes the maximum occurrence of failure range.

As shown in Figure 5.8, the failure rate is plotted to understand the working age of a component. Similar to Figure 5.7, three parameters are considered in Figure 5.8. The plot shows the occurrence of failure based on the availability of components. It shows the curve start developing at the end of the process. Hence, many possibilities can cause failure. Since the component considered is the control valve, the major failure can occur when the gate valve is open 80% or performing at the highest rate of ow. Hence, Figure 5.8 visualizes the rate of failure based on the hazard function (λ).

Figure 5.9 shows the predicted rate of failure ($f(t)$) plot. This plot shows the prediction based on reliability and its depending other parameters (i.e., η, θ, and λ).

5.5 CONCLUSION

This paper focuses on dragging attention to get introduced to the latest trend in Big Data science. This research shows the importance of Big Data in today's industry architecture, where digitalization is a new fashion. Big Data is not only important in the computational application, but also important in hardcore industries such as process plants, oil drilling, and many other industries. Hence, to develop a bridge between Big Data and the process industry, this research combines with a case study to develop a relationship.

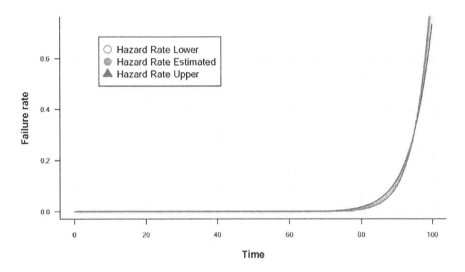

FIGURE 5.8   Failure rate plot vs time to visualize the (λ) values for failure occurrence.

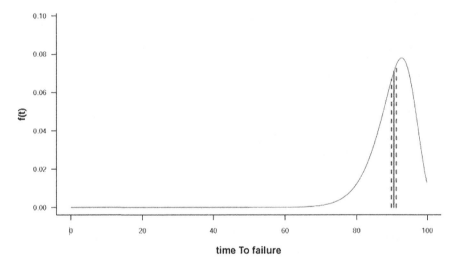

FIGURE 5.9   Estimated time of failure occurrence while operation.

The reliability prediction of a process plant is required to understand the system's actions in a real-time environment to ensure the application's long-term survival. Visualization of survival or risk rate is plotted on the bathtub curve for system failure (i.e., control valve) for reliability models. Therefore, the Reliability, Availability, and Maintainability (RAM) forecast, along with the Probability Risk Assessment, can be visualized on the bathtub curve. This will provide a visualization of the simulation for accuracy and comprehension of the instrument's working age and will help on-site engineers take the right step at the appropriate points.

REFERENCES

[1] Davis, J., Edgar, T., Porter, J., Bernaden, J., and Sarli, M. (2012) "Smart manufacturing, manufacturing intelligence, and demand-dynamic performance", *Comput Chem Eng* 47, 145–156.

[2] Collins, F. S., Morgan, M., and Patrinos, A. (2003) "The human genome project: lessons from large-scale biology", *Science* 300(5617), 286–290.

[3] Bickel, P. J., Brown, J. B., Huang, H., and Li, Q. (2009) "An overview of recent developments in genomics and associated statistical methods", *Philos Trans R Soc A: Math Phys Eng Sci* 367(1906), 4313–4337.

[4] Ebbers, M. "Things to know about big data in motion", 5. https://scholar.google.com/scholar?hl=en&as_sdt=0%2C5&q=Ebbers%2C+M.+%E2%80%9CThings+to+know+about+big+data+in+motion%E2%80%9D%2C+5&btnG=

[5] Jacobs, A. (2009) "The pathologies of big data", *Queue* 7(6), 10–19.

[6] Madden, S. (2012) "From databases to big data", *IEEE Internet Comput* 16(3), 4–6.

[7] Wu, X., Zhu, X., Wu, G.-Q., and Ding, W. (2013) "Data mining with big data", *IEEE Trans Knowl Data Eng* 26(1), 97–107.

[8] Gantz, J., and Reinsel, D. (2011) "Extracting value from chaos", *IDC iview* 1142 (2011), 1–12.

[9] Shewhart, W. A. (1931) *Economic Control of the Quality of the Manufactured Product*, London: Macmillan and Co Ltd.

[10] Reis, M. S., and Gins, G. (2017) "Industrial process monitoring in the big data/industry 4.0 era: From detection to diagnosis, to prognosis", *Processes* 5(3), 35.

[11] Weese, M., Martinez, W., Megahed, F. M., and Allison Jones-Farmer, L. (2016) "Statistical learning methods applied to process monitoring: an overview and perspective", *J Qual Technol* 48(1), 4–24.

[12] Qin, S. J. (2012) "Survey on data-driven industrial process monitoring and diagnosis", *Ann Rev Control* 36(2), 220–234.

[13] Ge, Z., Song, Z., and Gao, F. (2013) "Review of recent research on data-based process monitoring", *Ind Eng Chem Res* 52(10), 3543–3562.

[14] Manyika, J. (2011) "Big Data: the next frontier for innovation, competition, and productivity", http://www.mckinsey.com/Insights/MGI/Research/TechnologyandInnovation BigdataThenextfrontierforinnovation.

[15] Anagnostopoulos, I., Zeadally, S., and Exposito, E. (2016) "Handling big data: research challenges and future directions", *J Supercomput* 72(4), 1494–1516.

[16] Banaee, H., Ahmed, M. U., and Lout, A. (2013) "Data mining for wearable sensors in health monitoring systems: a review of recent trends and challenges", *Sensors* 13(12), 17472–17500.

[17] Hashem, I. A. T., Yaqoob, I., Anuar, N. B., Mokhtar, S., Gani, A., and Khan, S. U. (2015) "The rise of\big data on cloud computing: review and open research issues", *Inf Syst* 47, 98–115.

[18] Manyika, J. (2011) "Big Data: the next frontier for innova-tion, competition, and productivity," http://www.mckinsey.com/Insights/MGI/Research/TechnologyandInnovation/BigdataThenextfrontierforinnovation.

[19] National Research Council. (2013) *Frontiers in Massive Data Analysis*, National Academies Press. https://scholar.google.com/scholar?hl=en&as_sdt=0%2C5&q=National+Research+Council.+%282013%29+Frontiers+in+Massive+Data+Analysis%2C+National+Academies+Press.&btnG=

[20] TATA Consultancy Services. (2013) *The Emerging Big Returns on Big Data*, A TCS 2013 Global Trend Study, Elsevier.

[21] Kumar, D., Klefsjo, B., and Kumar, U. (1992) "Reliability analysis of power transmission cables of electric mine loaders using the proportional hazards model", *Reliab Eng Syst Safety* 37(3), 217–222.

[22] Billinton, R., and Allan, R. N. (1992) *Reliability Evaluation of Engineering Systems-Concepts and Techniques(Book)*, p. 620, New York: Plenum Press.

[23] Mathur, N., Asirvadam, V. S., Abd Aziz, A., and Ibrahim, A. (2020) Control valve life cycle prediction and E ect of valve stiction in reliability analysis, In *2020 16th IEEE International Colloquium on Signal Processing & Its Applications (CSPA)*, IEEE, pp. 46–51.

[24] Mathur, N., Asirvadam, V. S., Abd Aziz, A., and Ibrahim, A. (2018) Visualizing and predicting reliability of control valves based on simulation, In *2018 IEEE Conference on Systems, Process and Control (ICSPC)*, IEEE, pp. 54–59.

[25] Desborough, L. (2001) "Increasing customer value of industrial control per-formance monitoring-Honeywell's experience", *Preprints CPC* 153–186.

[26] Brasio, A. S., Romanenko, A., and Fernandes, N. C. (2014) "Modeling, detection and quantification, and compensation of stiction in control loops: the state of the art", *Ind Eng Chem Res* 53(39), 15020–15040.

[27] Choudhury, M. S., Thornhill, N. F., and Shah, S. L. (2005) "Modelling valve stiction", *Control Eng Pract* 13(5), 641–658.

[28] "Control Valve Dynamic Specification", Emerson.com, 1988. [Online]. Available: https://www.emerson.com/documents/automation/manuals-guides-control-valve-dynamic-specification-pss-en-67756.pdf.

[29] Controls, F. (1977) *Control Valve Handbook*, Marshalltown, IA: Sher Controls International Inc..

[30] Porter, L. J., and Parker, A. J. (1993) "Total quality management|the critical success factors", *Total Quality Manage* 4(1), 13–22.

6 Tracking or Maintenance of Device Parts

Tracking IoT-Based Devices Parts from Manufacture to End of Life by Leveraging Big Data in Blockchain Environment

S. P. Gayathri
The Gandhigram Rural Institute (Deemed to be University)

S. Vijayalakshmi
CHRIST (Deemed to be University)

Siva Shankar Ramasamy
Chiang Mai University

CONTENTS

DOI: 10.1201/9781003081180-6

6.1   INTRODUCTION

The entire world is moving toward the era of digital innovations. Without any doubt, the Internet is the initiation of innovation of digital ledgers, which are used in the blockchain. In every application, the blockchain and IoT concepts are increased in this digital era. In this chapter, the title "Tracking or Maintenance of Device Parts: Tracking IoT-Based Devices Parts from Manufacture to End of Life by Leveraging Bigdata in Blockchain Environment" clearly states that how an object or data in the blockchain and IoT plays an important role in manufacturing, education, immigration, agriculture, tourism, trade, cross-border commerce, e-commerce, wildlife, harbor, hospitals, heritages, counter-terrorism, police administration, management, Fintech, governance, etc. The term "blockchain" can lead each object or transaction between nodes to leave a transparent e-footmark through the network [1]. The Internet of things defines the objects or components connected in a network to do a specific task assigned [2]. This chapter discusses the possibilities of both blockchain and IoT on the objects or data throughout many known fields that impact human lives directly and indirectly. Day by day the count on the application is increasing exponentially in the real-time scenario. Blockchain and IoT are the perfect combination to build an architecture for innovative agriculture techniques such as the material conscious and information network (MCIN) model [3] for smart agriculture, which is different from the current vertical architecture. This blockchain-based model involves production, management, and commerce. This model uses enterprises-cum-individual personalized portals as the carriers which are linked precisely through a peer-to-peer network called six-degrees-of-separation blockchain. The authors want to symbolize a self-organized, open, and ecological operational system. This model also includes active, personalized consumption; direct, centralized distribution; and distributed production for smart agriculture. Discussions will be ongoing regarding the integration of the blockchain with the IoT with highlighting the significance and limitations. The authors [4] think that moving the IoT system into the decentralized path may be a smart decision because the blockchain is a powerful technology that can decentralize computation and management processes to solve issues related to IoT security issues. Another work presented integration [5] between IoT and blockchain based on secured health-related platforms to mitigate nurses' shortage. In that work, blockchain was used in the proposed operational framework to store and validate patients' records relating the nurse availability. Prototypical implementation on the proposed healthcare service was provided with all technical requirements to make sure blockchain and IoT shall give a successful result for healthcare systems. Another novel blockchain-based IoT model [6] was providing advanced security and privacy on hospital-related properties to the current IoT-based remote patient monitoring system. By using blockchain in the IoT-based model, the authors tried to eliminate challenges and improve the security of healthcare data in the hospital environment.

The model provides reliable data communication over the closed network and stores data over the cloud. The storage system used an ARX encryption scheme, which is an advanced and lightweight cryptographic technique. The authors introduced the concept of ring signatures to provide privacy properties such as signer's anonymity and signature correctness. In the same work, the authors used a double encryption scheme to make the symmetric key more secure over the network, used the Diffie–Hellman key exchange technique toward the blockchain-based network which protects the public key from any intruder.

The article that is related to supply chain [7] introduced the history and background, in terms of information science, for the supply chain background. The relevant technologies have been applied to improve the potential coverage for the purposes such as lowering cost, facilitating its security, and convenience. This work provides a comprehensive relative research work and industrial cases related to several companies. This work also illustrates the IoT enablement and security issues of the current digital supply chain system. The Combination of the existing blockchain's role and IoT-enabling technologies concludes a potential side for supply chain and its security issues.

Another article [8] discusses the use of blockchain in the education system, where the documents shall be traced through digital ledgers and electronic tracing such as students, teachers, and e-classrooms. The discussions from that article are mainly focused on keeping track of students and student-related records and transcripts, reducing manual and paperwork. Another article [9] supports the education system through IoT by providing the idea to develop learning kit training for the students. This concept showcases the impact that IoT and blockchain combination in the education sector, where the kit based on IoT consists of three parts, namely "brain", "muscle", and "cloud". The article used Raspberry Pi as the "brain" of operation that will be interacting with the Caiser Cloud platform. A test box is considered a "muscle" to provide simple data input. Then, the input from the test box will be sent, stored, monitored, and displayed on a platform created using Xojo software. The results of this article showed that the learning kit successfully interacts with the Caiser Cloud platform. The combination of blockchain and IoT shall create many tools that can be used as training tools for the education sector and the development sector as well.

Another article [10] discusses a survey presenting IoT security solutions after observing the lack of publicly available IoT datasets that were used by the research and practitioner communities. After the study on the potentially sensitive nature of IoT datasets, these authors claim that there is a need to develop a standard for sharing IoT datasets among the research and practitioner communities and other relevant stakeholders. Thus, the article posits the potential for blockchain technology in facilitating secure sharing of IoT datasets and integrity on IoT datasets and securing IoT systems.

Another paper [11] is related to a survey on authentication and privacy in IoT using blockchain during forest fire. This paper used blockchain to construct a distributed ledger of transactions that cannot be tampered with by any third party. The blockchain forms by cryptographically linked blocks of transactions with public-key cryptography and private-key cryptography which can be generated by officials and used by a closed group of people or government-related officials.

Having a glance at various fields, this chapter will discuss the possibility of using blockchain and IoT in various fields which may lead the entire world to the next generation. This chapter may not technically handle the process, but use the technical knowledge in various fields for increasing the interests of the readers or researchers. The previous discussions includes any object produced in the manufacturing sector shall be monitored and tracked until it reaches the retail shops. Then, objects or parcels or animals transferred by ship through harbor management shall use blockchain- and IoT-based technologies. The, students, teachers, certificates information shall be converted into digital nodes and make use of monitoring them and implying tracking them through IoT in the education sector. Passports and visas for the people residing outside the countries shall be monitored and tracked by embassies. The police officials, crime records, and criminals can be monitored and tracked by the police sector. Monuments, statues, and national treasures and heritages can be monitored and tracked by the blockchain and IoT tools. The important discussion in this chapter is implying blockchain and IoT on weapons in the future era to find out the right reasons to attack and fix the peace by reducing the usage of weapons. Then, the doctors, patients, and health records can be monitored and tracked by the health sector and a global level of improvement can be attained through blockchain and IoT tools. COVID patients can be monitored and tracked throughout the pandemic period through blockchain and IoT. In the main part of the world, the forests and wildlife shall be monitored and traced for the welfare of maintaining the balance of the entire species. An innovative and new side of business, financial technology (Fintech), shall be monitored through blockchain and improved for next-generation finance. After the COVID era, the world needs more support from cross-border commerce, so we can watch and trace through blockchain and IoT, which will help us to attain the peak of the performance on commerce. This chapter had a dream on the decentralization of Asia through blockchain and IoT, like connecting all the smart cities in Asia for attaining the fast growth and smartest way of living through the countries, by exchanging the culture, language, business, education, health system, and so on.

6.2   OBJECT TRACKING IN THE MANUFACTURING FIELD

In a manufacturing environment, the blockchain will provide an ID for every device; a public key will be generated from the manufacturer and a private key will be given to the consumer. Consider an object or data is being transported from India to Thailand. The Indian manufacturer will generate a public key, and the private key will be sent through another channel to the consumer in Thailand. The public key will be encrypted and the decryption key will be given to the consumer, by which we can access or get the product (Figure 6.1). In the blockchain, keys will be encrypted and decrypted to finish the complete process. This process cycle makes sure the transaction is being saved in adjacent nodes. Every important object's, making ID, manufacturing date, making cost, quality, manufacturer, supplier, client, and end users information will create a Bigdata, but blockchain will monitor and maintain the track record efficiently through the trace ID, public key, and private key given to the end user of the client.

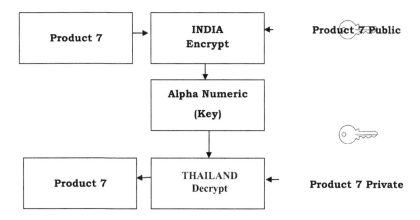

FIGURE 6.1   Object's information is encrypted and decrypted in a blockchain.

6.3   HARBOR – SHIPPING

Considering every ship as a network node, and every container as a node in a ship, we can use the tracking process in shipment. For every shipment, there is a sender and a receiver and the object is also mentioned in the receipt, so this shipment will have data such as the sender, the receiver, the product type, permission, weight, approximate price, date of sending, approximate date of receiving, and purpose. These details may look very simple, but imagine a harbor permitting passage for approximately 100 ships per day. Assume that a ship can transport approximately 500 containers, and if each container is carrying ten different goats, in such a situation, we are entering Big Data into blockchain technology. The manufacturing industry requires a chain of process and properties (Figure 6.2). The Entire process will be converted in distributed transparent digital ledgers in the blockchain technology will have a distributed participant's copy, the secure encrypted keys, validated record with process possibility, identification of all the involving participants, the validity of each product's record, traveling distance, transaction bodies involved in them, currencies involved, etc. The proper time-stamping of each record will help the shipping

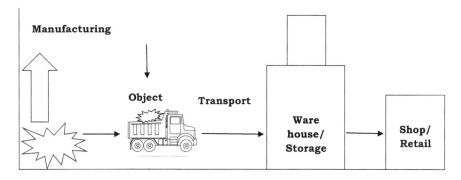

FIGURE 6.2   The raw material starts from the collection point to a shop as a product.

company to provide the demand area, the manufacturer, harbor management, shipping companies, the government involved, the policies of cross-border commerce, and the quality of the product as well. The customer feedback can also be added to a node if permitted in the blockchain.

6.4 EDUCATION – TRACKING

No doubt technology reaches its peak curve with the foundation of education. Nowadays, the education is also being reformed by technology and available in our hands. The beauty of education in technology and technology in education is helping hand in hand with improving the accessibility and quality through the world of education. Figure 6.3 shows that a university or institution shall have all the digital records of the teachers, students, mark statements, and alumni details in the organized network. Blockchains shall be used to connect them as adjacent nodes, by which all the transactions or actions of every node shall have a trace through the network. As sustainable development goals, education to all will be possible when we

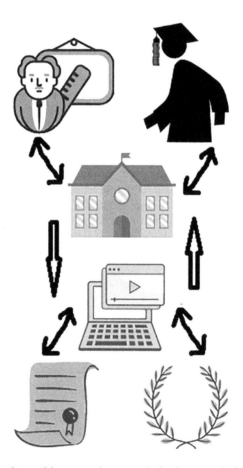

FIGURE 6.3 University tracking on student records for future analysis.

imply blockchain and IoT on both technology and education within the next 5 years. The certainty in education requires keeping track of teachers, students, and related records such as classes, materials, and communication. The transcripts after examinations and results are more significant than we imagine. Every student's record, the subjects he study, the teachers who handle those subjects, their questions, answers, assignments, materials, timetable, attendance, feedback, classroom, course feedback, parents' feedback, and end users' feedback will create a Bigdata, but by using blockchain's digital ledger, we can efficiently keep the trace ID, and that can be used for ranking and regulatory committee in future. Centralizing and distributing the records may give a world-class comparison and competition as well.

Reduction in manual and paperwork [8] in the education field will save time, money, and efforts of many people directly and indirectly. Pursuing the graduation rates is increasing rationally, but obtaining the original certificates and acknowledgment certificates for applying for the jobs and promotions in career leads to heavy stress for the students and candidates throughout the world. This step will improve the quality of the education sector as well. The unique IDs, passport numbers, mobile numbers, and emails that are linked with a unique ID shall be linked with every institution's tracking network and shall be monitored by blockchain (Figure 6.4). The discussion of the tracking of the transcript and the student status can also be used for educational institution rankings, the nation's education process and policies, United Nations surveys, education visa, employment visa, global employments, and research and development throughout the world [12].

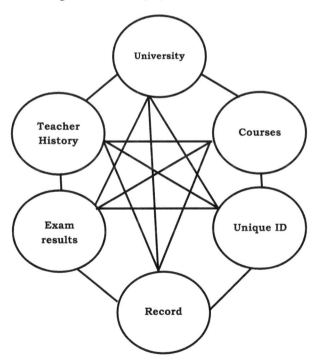

FIGURE 6.4 The node construction of the blockchain possibilities in the education sector.

The certificates, personnel, and employment are very important to the welfare of any nation or organization. There shall be a check point to avoid any criminal activities or criminals or criminal methods inside any organization. The criminal check can be done through blockchain by confirming the nodes connected in the network traces on reversible methods. Proper education and the tracking of the students as well as teachers will reduce terrorism and increase peace and harmony throughout the world. The biggest mistakes in history are done by educated people, so the educated people can be monitored more. At the same time, the selection and continuation of higher studies can be monitored under government bodies. As per United Nations's sustainable development goals, education to all will be possible when we imply blockchain and IoT on both technology and education within the next 5 years. The certainty in education requires keeping track of teachers, students, and related records such as classes, materials, and communication. The transcripts after examinations and results are more significant than we imagine. Reduction in manual and paperwork in the education field will save time, money, and efforts of many people directly and indirectly. Pursuing the graduation rates are increasing rationally. Obtaining the original certificates, a while applying for the jobs and promotions in career leads to heavy stress for the students and candidates in India. In recent years many graduates are facing troubles in obtaining equivalent certificates for their programs from various universities around the world. Implementation of Blockchain will reveal the association between the course outcomes, program outcomes for the clarity in studies as well job requirements.

6.5 EMBASSY – PASSPORTS

Tracing of records for non-resident citizens of every country can be effectively done through blockchain. This blockchain concept can be applied in the context of every embassy or consulate as network nodes and every non-resident immigrant as a node. Blockchain can use the tracing process through entries and physical presence dates, times, and health records when the citizens are from outside of the country (Figure 6.5). For every visa extension, there will be an address for stay and work, and the consulate visiting dates and extension dates are also mentioned in the receipt, so the visa extension will have data such as the date of applying, date extended, the passport number, work permit number, visa type, days extended, place of stay, place of work, dependent visa, and amount paid for extension.

These details may be crucial for every citizen from outside of the country. We get a lot of complaints from the people who work in abroad related to their stay and job nature and salary.. If a consulate is processing 1000 immigrant visas per day, approximately 10,000 basic details will be verified and updated in the global network. This Big Data leads blockchain technology to a thorough update in the tracking process. The properties of distributed ledger in the blockchain technology will have a distributed immigrant's copy, the secure encrypted keys, validated record with the reversible and irreversible process, identification of all the involving immigrants, visa verifying officer, consulate general, policemen, working organization, the place where the immigrants stay, and finally all the transactions bodies and currencies involved in them. By a proper visa process and immigration process, the future can

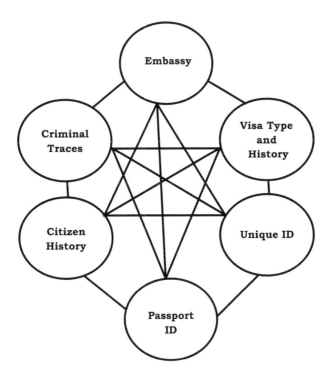

FIGURE 6.5 The node construction of embassy relations to passport.

identify the issues faced in the past times, recovering issues in the present stages and avoiding mistakes in the future as well. Every immigrant's visa record, places he traveled and stayed, the hotel owner or the friend he met, his feedback on the place, feedback on him and his actions, the money he spends, the actions he did, the materials he purchased, time and date, etc., will create a Bigdata, but by using blockchain's digital ledger, we can efficiently keep the trace ID, and that can be used for counter-terrorism and regulate the visa conditions and methods in future. Centralizing and distributing the records may give world-class tourism data and shared human race traces as well. We can counter the terrorism by sharing the red-list passport to the consulates throughout the world, by which if any person comes to cross an issue based on passport holder or an immigrant, then local police, national police, or Interpol can also be involved issuing arrest warrant with the complete data that existed. If a person is trying to hide, a person is giving wrong information, or a person is suspected of unwanted movement, he shall be arrested and intimated by the local police or regional army command, and the information shall be passed to the consulate, embassy, foreign minister, or home minister of that particular city.

6.6 POLICE STATIONS – CRIMINAL RECORDS

Tracing criminal records for the resident of crowded cities in every country is a tough task, but a required task to increase the global peace index. This can be effectively done by connecting the records in the blockchain network and updating through devices controlled by IoT, throughout the cities. The blockchain concept can be applied to social security numbers or unique IDs given by every country. The unique ID won't change, so the family members, neighbors, and residents of the city can be organized based on the activity as well as movements. Every criminal record can have at least one clear government proof and contact number. Blockchains can use the tracing process through entries and physical presence dates to the police station or the respected courts. The time, vehicle, mobile number, and health records can also be updated with a priority-based storage system. When the criminal or a suspected citizen is outside of the country, the embassy shall give the information or updated information to the concerned country's request (Figure 6.6).

Its very important to know, monitor and share the change of the address, stay, work, contact information shall be traced. The criminal networks are done in various methods such as regular meetings, telephone calls, and third-party communication. The record and suspected activities' dates, places, connections between the criminal, and money transfer can be analyzed to find out the possibilities. There shall be no exception given on updating the criminal records because they are crucial for every citizen inside the country and outside of the country. The police station used to get various complaints from the people for simple reasons that may not be avoided as a small issue, many criminal activities nowadays have a very simple reason behind them. The police shall visit the place based on the complaint, shall inquire to both parties, and shall address the direct witness as well as the indirect witness. The police station is the local source for every citizen, where they are responsible for keeping

FIGURE 6.6   The node construction of crime system relations to police administration.

peace and order within a certain square kilometer. Any irrelevant movement, issues, functions, festivals, rallies, and gathering shall be observed through CCTV cameras throughout the city. The records are converted into Big Data, which leads the block-chain technology to make a thorough update in the tracking process or the missing process. The citizen or criminal records on the distributed electronic ledger in the blockchain technology can be distributed to the index-based central hubs that can be used by higher intelligence and government officials. The distributed e-ledger copy and the secure encrypted keys can be maintained by a highly secured wallet in every country. Highly motivated students or professors can be used in validating the records or ledgers in a decent period to find out the backpropagation process through the records. Identification of commonalities can be immediately informed to district- or state-level officials, data verification officers, or organizations related to the person. If the suspect or criminal is outside the country, the information shall be passed to the Consulate General, visa officials, and police stations near the place where the immigrant stays or works, and people related next to him. We can counter the terrorism by sharing and tracing the criminal list and red-list passports to the consulates and legal intelligence agencies executed by the governments throughout the world. Any person who comes across an issue abroad shall be interrogated by local police, the national police official, or Interpol if it is required. This can be extended to issue an arrest warrant with the complete data that existed. If a person is trying to hide or modifying any important information or providing wrong information, or a person is suspected of being part of banned movements, he shall be arrested and intimated by the local police or regional army command, and the information shall be passed to the consulate, embassy, foreign minister, or home minister of that citizen. The global records shall be updated up to the local records whenever the cloud is allowing access to the private-key holders.

6.7 MONUMENTS AND HERITAGES – TRACKING

India is a subcontinent with unlimited heritages within it. India itself can balance the entire world with the count of heritage sites and monuments. India has a lot of diversity, and that becomes the positive side of having various types of heritage sites and monuments. Every state in India is having unique cultures and traditions. The heritage sites in the places are the roots of the tradition and customs being followed there. Languages are the roots of the tradition and archeological sites are the evidence for every culture. Blockchain shall be implemented through the entire Asia connecting heritage sites, and a comparison shall be done on transactions of language, culture, communication, and historical traces also (Figure 6.8). Many common habits, beliefs, traditions, and festivals are being followed in countries such as India, Thailand, Myanmar, Vietnam, Cambodia, Laos, Sri Lanka, Indonesia, Malaysia, Pakistan, Bangladesh, Bhutan, Nepal, Tibet, China, Japan, the Bali islands, Singapore, the Philippines, Taiwan, and South Korea. All these shall be preserved by sharing the cultures by inviting each other on special occasions. These communications keep those traditions alive. The world will be alive when we have diverse cultures and traditions.

Let us consider that every culture has a unique touch and root in their history. Those roots were considered as data in the node, and these shall be encrypted, stored, and transferred to the related nodes nearby. The war or disputes between the countries are happened because of doubt and greed on the adjacent countries' wealth and to occupy the historical treasures or to erase the historical treasures. If the traces are stored digitally and spread throughout the world, the countries can have more evidence or more movements by enlightened tourism and known date of a particular place or a monument. Languages are spread by traveling and business and cultural exchanges. For example, the Tamil language (Figure 6.7) has a trace through countries such as Sri Lanka, Singapore, Malaysia, Indonesia, Cambodia, Vietnam, Thailand, Myanmar, and South Korea. Let us take a root word as a node and we can find the traces within the languages, cultures, traditions, and festivals are being followed in different names through southern Asia. In this case, encryption happens naturally in the people's regional languages, and decryption happens when the words have similar phonetics or scripts or expressions or meaning. Scripts, languages, images, evidence, stone scripts, metal evidence, palm leaf scripts, cotton scripts, phonetics, poems, stories, histories, places, routes, geological information, and architecture types will form a Bigdata again, and when we use a unique ID for the objects, they can be digitally stored and can be traced all the time. The human race has a unique beauty that we enjoy the commonalities as well as the diversities. Blockchain and IoT shall keep the information alive and pass it for the next generations.

This step will improve the quality of the education sector as well. The unique IDs, passport numbers, mobile numbers, and emails that are linked with a unique ID shall be linked with every institution's tracking network and shall be monitored by

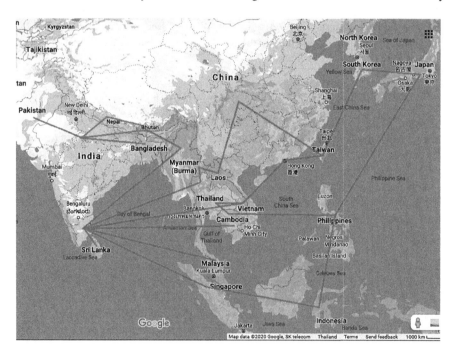

FIGURE 6.7 Countries possessing some common languages, words, culture, traditions, festivals, beliefs, etc.

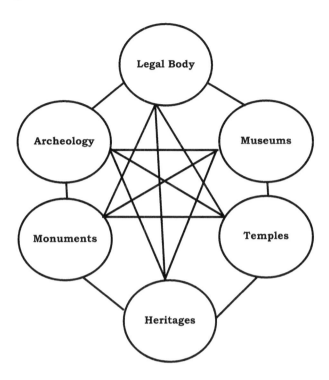

FIGURE 6.8 The node construction of blockchain possibility in heritages and monuments.

blockchain [12]. The discussion of tracking the transcript and the student status shall also use for educational institution rankings, nations education process and policies, United Nations surveys, global employments, and research and development through the world. There shall not be any criminal activities or criminals or criminal methods be implied inside an organization [13]. The criminal check shall be done through blockchain by confirming the nodes connected in the network traces on reversible methods. Proper education and the tracking of the students as well as teachers will reduce terrorism and increase peace and harmony throughout the world. The biggest mistakes in history are done by educated people, so the educated people shall be monitored more [12]. At the same time, the selection and continuation of higher studies can be monitored under government bodies.

6.8 WEAPONS AND WAR

Every country in this world has police and army services to uphold the law and security, and for that, they need weapons.

We cannot link all the weapons through the network, but we can link important weapons in the control of some ministries, which can be responsible for operating those weapons, such as fighter jet, missiles, submarines, warships, important tankers, Army commuters and so on. Each important weapon can be considered as a node and connected through a private blockchain, and we can see the process of movement and actions in them (Figure 6.9). This updation may give a lot of clarity through the

FIGURE 6.9 The node construction of blockchain in healthcare systems.

ministry, the nation, the subcontinents. This transparent link will flow through the entire world. The weapon trade will be organized and legalized without corrupted or untraceable money involved in it. The trace on underground weapon dealing will stop the battles between ethnic groups and the involvement of multiple nations as well. Illegal weapons such as chemical weapons and biological weapons will also be stopped by these blockchain digital ledger certifications. The responsible countries in the world shall start this process, and it will be followed and become a practice to every nation to provide their support to the peace and harmony through the communities.

6.9 MEDICINE – DOCTORS AND PATIENTS

The blockchain can be applied in the medical field having doctors and patients as the two different ends. Considering the hospitals as the networks, the medical practitioners, nurses, staff, and patients are connected with IoT-based ID cards into the network (Figure 6.10). Let us consider that a node is mobile in this process. Blockchain can do tracing the patient's movement, entries, and physical presence dates, times, and health records when the citizens are outside the country. Every patient will have his history, and that history can also be added as additional notes in the network. Every medicine he took, every test he attended, availability of doctor, every doctor he visited, or every nurse who gave him an injection or a tablet or a suggestion shall be recorded and transmitted through the network. This may show the clarity in the treatment, the transparency in the hospital, or the medical system they follow. Medical

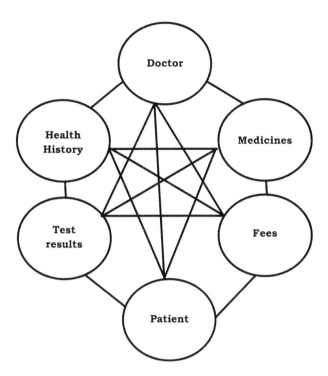

FIGURE 6.10 The node construction of blockchain in healthcare systems.

methods such as allopathy, naturopathy, acupuncture, Siddha, Ayurveda, and Yunani shall follow the same. Blockchain can easily identify the medical method, practices, pharmaceutical company, organizations' involvement, political influence involvement, the price, and the quality difference between a branded company and local companies. Then, we can have quality as well as inexpensive medical methods throughout the world.

A hospital is not only running through pharmaceutical-based medicines. Many hospitals are focusing nowadays on giving a multi-specialty method of testing the human body, blood, bone, tissues, and skin, magnetic resonance imaging, X-ray, urine test, stool test, testing of neurology issues and birth issues, psychological analysis, mental stress test, general healthcare, etc. All the above-mentioned parameters are considered nodes in the hospital network. One node will surely connect any other node in related network. This process will create the connected network. Any node or network will be associated one to another directly or indirectly.

6.10 COVID PANDEMIC

In recent months, the entire world is facing a COVID-19 pandemic, which requires a tracing method from a patient to unknown members. The use of blockchain can solve this problem a bit easier. Every patient can be considered as a node, and the other nodes can find the possibility of connection and the possibility of passing the data, in this case the virus. The place of the patient, the hospital where he is taking treatment, his family members, friends, colleagues, and the people with whom he can meet are all considered as positive or connecting nodes. Considering a date, time, work schedule, and any other possible parameter can be associated to predict these connections in the network.

6.11 FORESTS AND WILDLIFE

The next important resource of India is the forest. Every forest consists of numerous natural resources such as trees, rivers, valuable animals, multiple living beings, and herbs. Every forest will have a different type of geographical features and soil.

Forests possess a lot of mineral properties, so using blockchain, the information of the forest can be designed as a network, and the parameter can be assigned to the network in which nodes represent, for example, borders, routes, valuable trees, valuable animals, endangered species, important national treasures, important national animal, bird, food cycle, climate, herbs, research unit, forest range unit, forest guard unit, check posts, officers in charge, staff in charge, tribal people, the political structure in the forest, resources that can be used from the forest, permission from the government to use the resources, forestation process, firefighting equipment, firefighting

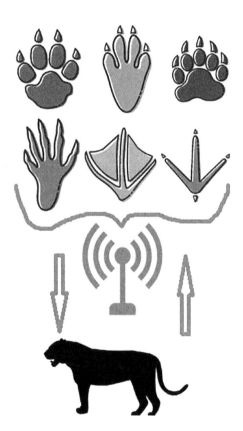

FIGURE 6.11   Wildlife into blockchain tracking through IoT.

units, nearby fire stations, biological and geological expert committee, universities which are using the forest for research and development, and political member associated with the forest. Counting, monitoring, and migration of the important animals can be done by IoT object recognition.

Blockchain can be utilized for keeping the trace alive and informing them to the higher officials and volunteers with a digital ledger (Figure 6.11). Volunteers related to forests can be linked into this network by extending the transparency of the livelihood and awareness to the public. Veterinary doctors may be activated in conservation camps with proper safety measures. Animal products can be legalized and monitored and tracked; for example, honey and leather can be handled by any organization, but national treasures such as "ivory", "tigerskin", and "peacock feather" can be properly preserved and sold through the government portal only. Crime on wild animals or regional heritage animals shall be treated as the highest crime, and a statement shall be passed to the public.

Blockchain and IoT can also be used to prevent forest fires and to alert communities nearby. Deforestation and soil erosion can also be identified and properly alarmed before the next level of collision is held in the forests. Pollution and using irrelevant materials in the forest can be identified by IoT, and the blockchain shall have the records related to the people living in the forests and officers who are working in the zones. Accidents shall be immediately reported and issues shall be solved by having IoT-tracked ambulance services in the forest zones or wildlife zones. Tourism shall be improved by accessing a safer building in the forests without disturbing the wild animal's path and livelihood. Tourist hubs shall be created, connected, and information related to those hubs shall be shared for a transparent network. Such network will provide transparent transportation, hotel booking, days visited, time visited, members visited, language used in certain tourist spot, traveling plan, safari, purchase of tickets, traveling, entry and exit place and time from the forests shall be stored as digital ledgers. This action may bring clarity in law or order through the tourist places. If Law and order is maintained transparently, India can have more tourists. The different languages in India, is also a barrier for the foreign tourists in India. Linking the important verbs between the highly used languages in India, will create mostly used block chain ever.

6.12   FINTECH THROUGH BLOCKCHAIN

People can know the transparent information from blockchain through their mobile or a system, which is open to the world. The implementation of blockchain in every transaction hesitation, but later when governments impose this as an act, then the process can make a big change. Building a Cross Border E-Commerce [14] through blockchain is based on taking the blockchain technology from normal finance networks into innovative digital platforms. There are two key costs identified to get affected by the technology. For every transaction and tracing, the two key costs are the cost on verification and the cost on networking all the nodes. The digital economy or digital Fintech shall be achieved by combining the ability to secure and verifying the status within the transitions that are valuable from a network perspective. This also requires the contribution of the resources to operate, scale, and secure a decentralized network. The Fintech marketplaces allow participants to join and operate joint investments in shared digital infrastructure and public utilities without assigning market power to a platform operator. This act is characterized by an increased competition, reduced barriers to the entry level, and a lower privacy risk.

6.13 CROSS-BORDER E-COMMERCE

Today's world is practicing the new normal lifestyle, and new e-commerce requires online transactions and traceable objects. Cross-border e-commerce shall be updated with IoT devices for storing data and connecting through blockchain digital ledgers. Cross-border e-commerce transactions between the Asian countries are improving on an unexpected high scale after the 5G implementation in China. After the India–ASEAN Meet in November 2020, Asia shall grow rapidly through digitalizing the entire commerce activities through the blockchain and digital currencies as well. For commerce activities, there will be a threshold level of amount, which shall be traced and monitored. The implementation of blockchain seems expensive, so the threshold level shall be fixed through financial and IT experts. The business and communication tracing serves every person, such as customers, middlemen, and merchants. The barriers such as language, cross-border transfer policies, and post-sales services are direct and indirect barriers in implementing the blockchain through Asia, but the trace remains forever, by which people can trace their products or purchase orders. Through the blockchain–implementation in B2B and B2C business model, the business will have the customers, clients, merchants, material supplier's, middle men's authentication, banking details and universal-level ID. This information shall be stored in Blockchain network and used for digital tracing for all the process through the business transactions. At the same time, it is important to know the storage level, and modifications and removing redundant data may be a challenge.

6.14 DECENTRALIZATION OF ASIA

Digital India and Digital Asia no longer a dream. It has the infrastructure such as connecting the local networks and digital data toward 5G and IoT through blockchain digital ledgers. South Asian countries such as South Korea, Thailand, India, Malaysia, Singapore, Indonesia, Myanmar, Laos, Vietnam, Cambodia, Bangladesh, Pakistan, and Sri Lanka shall use this pandemic situation to adopt the faster and transparent digital ledger system on business and finance. This shows their honesty in the world trade segment. Any organization or countries involving in terrorism won't join into clusters of the Blockchain. They may be aware of the currency used for terrorism, fund, sender, receiver, reason, purchase history, people involved, device involved, networks involved and time will be traced. There are lot of proof on few countries, who are involving on terrorism activities. They may avoid implementation of Blockchain. But by the global pressure from the world's leading organizations, neighboring countries, bring the terrorism to an end. Supporting terrorism is not only ending with attacking other countries. Supporting terrorism extends activities such as allowing and sharing unwanted products such as opium, hemp, illegal weapons, illegal money, child trafficking, slave system, illegal prostitution, unauthorized medicines, and virus research on Bio-war fare.

6.15 CONCLUSION

Blockchain and IoT will be the perfect combination to trace any object that can be digitally identified and monitored. Digital data can travel the entire world via a traceable network. Big data or the digital objects connected with the Internet require maintenance to secure the transactions. Blockchain technology leaves an electronic ledger by providing a key for transactions, which reliably make the items or data track back or monitor instantly. Each transaction contains a lot of data, and those data are stored in multiple links or nodes. Even this task provides the positivity and transparency of any transaction combining blockchain and IoT devices for the current e-world. All the above-said properties will be used in the fields of medicine, agriculture, education, wildlife, nature, business, counter-terrorism, research and development, sustainable development, governance, cross-border commerce, market, trading, etc. Tracking IoT devices through blockchain methods is part of digital innovations, which are leading the way in the post-pandemic era. One way or another, 60% population of this world will be benefitted from the entire activities going around the blockchain and IoT concepts. Any object or device will have several tracing properties, and those properties shall be considered as nodes and the communication between the nodes will be stored in multiple copies transparently. If any organization implements blockchain concepts in their IoT devices, the consumers and the company shall improve their quality and consistency to the next level. The chapter discussed the possibilities of objects, properties of blockchain in those objects or data, and tracking the IoT devices. The entire world should try such future technologies or digital innovation for their successors.

REFERENCES

[1] Haber, S., and Scott, S. W. (1991) "How to time-stamp a digital document", *J Cryptol* 3(2), 99–111.
[2] Narayanan, A., Bonneau, J., Felten, E., Miller, A., and Goldfeder, S. (2016). *Bitcoin and Cryptocurrency Technologies: A Comprehensive Introduction*, Princeton: Princeton University Press.
[3] Gu, X., Chai, Y., Liu, Y., Shen, J., Huang, Y., and Nan, Y. (2017) "An MCIN-based architecture of smart agriculture", *Int J Crowd Sci* 1(3), 237–248. doi: 10.1108/IJCS-08-2017–0017.
[4] Atlam, H. F., Alenezi, A., Alassafi, M. O., and Wills, G. B. (2018) Blockchain with the internet of things: benefits, challenges, and future directions", *Int J Intell Syst Appl* 6(2018), 40–48.
[5] El-dosuky, M. A., and Eladl G. H. (2020) "Using IoT and blockchain for healthcare enhancement", In (Rocha, Á., Adeli, H., Reis, L., Costanzo, S., Orovic, I., and Moreira, F., eds.), *Trends and Innovations in Information Systems and Technologies. WorldCIST. (2020) Advances in Intelligent Systems and Computing*, vol. 1159, Cham: Springer. https://doi.org/10.1007/978-3-030-45688-7_63.

[6] Srivastava, G., Crichigno, J., and Dhar, S. (2019) A light and secure healthcare block-chain for IoT medical devices, In *2019 IEEE Canadian Conference of Electrical and Computer Engineering (CCECE)*, Edmonton, AB, pp. 1–5, doi: 10.1109/CCECE. 2019.8861593.

[7] Zhang, H., and Sakurai, K. (2020) "Blockchain for IoT-based digital supply chain: a sur-vey", In (Barolli, L., Okada, Y., and Amato, F., eds.), *Advances in Internet, Data and Web Technologies*. EIDWT (2020) *Lecture Notes on Data Engineering and Communications Technologies*, vol. 47, Cham: Springer. https://doi.org/10.1007/978-3-030-39746-3_57.

[8] Hashmani, M. A., Junejo, A. Z., Alabdulatif, A. A., and Adil, S. H. (2020) Blockchain in education – trackability and traceability, In *International Conference on Computational Intelligence*, pp. 40–44.

[9] Kamal, N., Md Saad, M. H., Kok, C. S., and Hussain, A. (2018) "Towards revolution-izing stem education via IoT and blockchain technology", *Int J Eng Technol* 7(4.11), 189–192.

[10] Banerjee, M., Lee, J., and Choo, K. K. R. (2018) "A blockchain future for internet of things security: a position paper", *Digital Commun Networks* 4, 149–160.

[11] Datta, S., Das, A. K., Kumar, A., and Sinha, D. (2019) Authentication and privacy preser-vation in IoT based forest fire detection by using blockchain – a review, In *International Conference on Internet of Things and Connected Technologies (ICIoTCT)*.

[12] https://www.education.wa.edu.au/dl/xlkqdr.

[13] https://www.unescap.org/sites/default/files/Agenda%20Item%202%28e%29_ Emerging%20technologies.pdf.

[14] Bibi, S., Siva Shankar, R., Nopasit, C., and Piang-Or, L., (2021) "Improving digital platforms and B2B2C strategies for cross border e-commerce", *Turk. Online J. Qual. Inq.*, 12(6), 5232–5244.

7 Jaya Process Optimization in IoT Applications

G. Revathy
Sastra University

K. Selvakumar
Annamalai University

T. Ganesh Kumar
Galgotias University

CONTENTS

DOI: 10.1201/9781003081180-7

7.1 INTRODUCTION

The outbound character of WMN is an optimistic inclination in which wireless networks. The crucial support of WMN slouch in its native gaffe limit alongside network failures, minimalism to milieu affirmative a network, and a broadband competence. WMN rigging elevated mobility appraise to erstwhile wireless networks since WMN has superior energy storage and power storage. The features of WMN are self-configuring and self-healing techniques in the course of the node malfunction or trail malfunction are slickly enhanced since, in WMN, a node can be full of zipping as the customer as well a server dependent on the request [1,2,4,5]. Aby lose of in the node of recovers routinely. In WMN, a node maneuver is as huge as a modem, barefaced packets on behalf of former nodes that are not in a wide variety of wireless communications. The WMN is distinguished by the pulsating self-organizational feature that makes it easy to introduce quickly, maintain problems, minimize costly costs, and deliver trustworthy services for desirable network power, bandwidth, versatility, and hardness [6,8]. The general idea is to build robust and efficient wireless broadband networks that need very insignificant front-run investments for cash-stricken operators, carriers, and others.

Figure 7.1 describes a typical WMN architecture. Two distinct clusters surround the mesh nodes. Figure 7.1 depicts simple figure mesh networking. When we go for sophisticated, there are various nodes in each cluster, and hence, when we want to

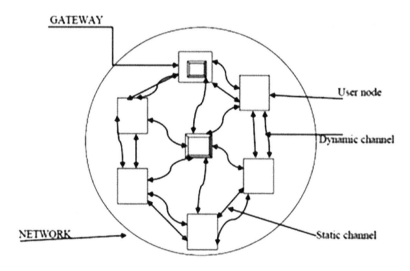

FIGURE 7.1 WMN architecture.

convey data from the initial place toward the target, we have a near recognized shortest path for broadcast. In this concurrence, we have realized a new technique of setting IoT sensors to each node and navigating fast using Jaya optimization.

7.1.1 Internet of Things (IoT)

IoT oration endows through an elementary, also superior, perception of IoT. Our IoT tutorial is anticipated for indispensability and proficiency. Our IoT assimilates all IoT principles such as prelude, features, augmentation and exasperation, ecological unit, decree framework, propose, province, biometric, security camera, door unlock system, and devices (Figure 7.2).

IoT reflects a refined categorization for optimization and research that agrees with computers, actuators, networking, electronic, cloud messaging, etc., to have sufficient markings for extension or facilities. Also, there is a dais, such as exhaust, which contains all the essential elements in the path surrounding us. For example, in a house, we can snip our home appliances, air conditioning, lighting, and so on, and all these items are handled on the same platform. The IoT harmonization has excellent accuracy, administration, and efficiency. Because we have a podium, we can shave the vehicle, follow its fuel meter, tempo altitude, and track the environment [6–8].

I might bring warmth in the room if there is a frequent forum, where all these items will package with each other. For starters, I would like the warmth to be 25 or 26 °C when I get home from my office, and my AC would set up 10 minutes before I get home according to my car location. This can be done in the IoT at some point.

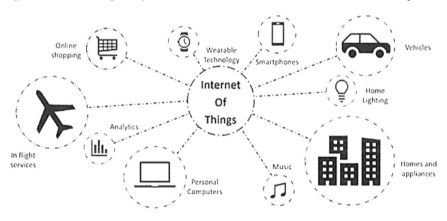

FIGURE 7.2 Applications of IoT.

FIGURE 7.3 Work flow of IoT.

7.2 WORKFLOW OF INTERNET OF THINGS (IOT)

For different IoT echo entities, the importance of IoT is diverse (architecture). Nonetheless, there is an equivalent essential discernment. The intact development of IoT begins with the electronics, including smartphones, digital watches, and electronic devices, which are gradually paired with the IoT podium. The lector gathers and dichotomizes data from all large instruments and transmits and woodworks the most stylish information resulting in devices (Figure 7.3).

7.3 IOT ATTRIBUTES

Connectivity, observing, incorporating, and constructive interaction are the most vital aspects of the IoT mechanism (Figure 7.4). Here are a number of them premeditated:

a. **Accessibility**: belongings of connectivity to expose a related connotation in the IoT plinth that can be used or the cloud in all its components. Consequently, it offers a high momentum message among the diplomacy and the cloud concerning IoT devices to make the announcement consistent, sheltered, and bidirectional [9].

b. **Analysis**: It synchronizes the unruffled data and uses them to gather successful market information after affecting all the associated possessions. We mark our organization as possessing a smart device if we boast a successful eventual data set from these possessions.

c. **Integration**: IoT switches up the various styles to facilitate the familiarity of addicted persons.

d. **Artificial information**: IoT formulates chic stuff and increases life by manipulating data. For the coffee machine itself classifies the coffee beans

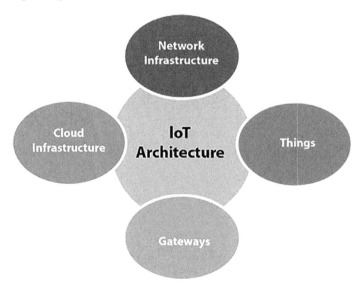

FIGURE 7.4   Features of IoT.

of your choosing from the supplier, whether we feature a coffee machine that has finishing beans available [10].

e. **Sensing**: The IoT technologies' sensor campaign acknowledges and tests any revolution in up-building and testimony. IoT equipment transmits flaccid networks to energetic networks. Except for sensors, an accurate or factual IoT environment cannot be recorded.

f. **Constructive engagement**: IoT shall express the technology, goods, or resources shared to abrupt assignments.

g. **Endpoint management**: The endpoint connotation of the whole IoT scheme is differentiated otherwise by the company's absolute wildcat attack. For, e.g., if a coffee apparatus itself orders the beans, what happens when it orders the beans from a supplier, and we are not home for a couple of days? It would escort the IoT machine to failure. Thus, endpoint management is essential.

7.4 IOT ARCHITECTURE

There is no unusual or customary agreement, universally established, on the IoT. IoT architecture is different and explained from its serviceable area. However, the knowledge in IoT architecture primarily consists of four virtual devices:

7.4.1 COMPONENTS OF IoT ARCHITECTURE

- Devices/Sensors
- Networks and Gateways
- Application layer cloud/organization
- Layer of Functionality.

There is an abundant stratum of IoT rigid leading the commonness and recital of IoT rudiments that bestow with the most profitable solution to the exchange endeavor and end-users. The IoT devise a straightforward mode to intend the sundry prerequisites of IoT, so to succor, it canister apportion services above the overtone and dole out the chucks for the future.

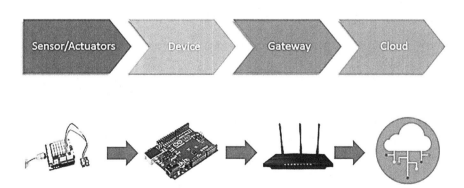

FIGURE 7.5 Stages of IoT solutions architecture.

Again are the essential intersections (layer) of IoT, which is aligned with the compulsive role of the IoT principle (Figure 7.5).

1. **Sensors/actuators**: Sensors or actuators are intellectual campaigns to produce, understand, and advance knowledge in the network grasp. Sensors or actuators may be enthusiastically enmeshed in a wide variety of frenetic or wireless devices. This encloses GPS, gyroscope, GPS, RFID, etc. Most of the sensors are linked through sensor gateways. The communication between sensors and actuators may be from a personal area network or a local area network [11].
2. **Data collection and gateways**: These sensors dictate fast gateways and ties for the redistribution of the data since the healthy statistical information is modeled. The link container falls in the shape of the Local Area Complex (Ethernet, LAN, WiFi, etc.) (WAN like GSM, 5G, etc....).
3. **Edge IT**: The edge is the software and hardware portal in the IoT architecture, which evaluates and preprocesses data from repositioned in the cloud. Should the data stored in the sensors and gates not be rehabilitated from its value as mentioned above, it does not redistribute the information used in the cloud.
4. **Data center/cloud**: Data center or cloud under the controls that test the building in the analytics, system management, and security tests. Following the cloud repository pads, the data are passed to end-users such as shopping, hospitals, emergency, climate, and electricity.

7.5 SMART ENERGY SYSTEM

IoT has a beneficial effect on energy management and control. This is referred to as the Smart Energy System. IoT apps track the number of energy management services used in home and commercial settings (Figure 7.6).

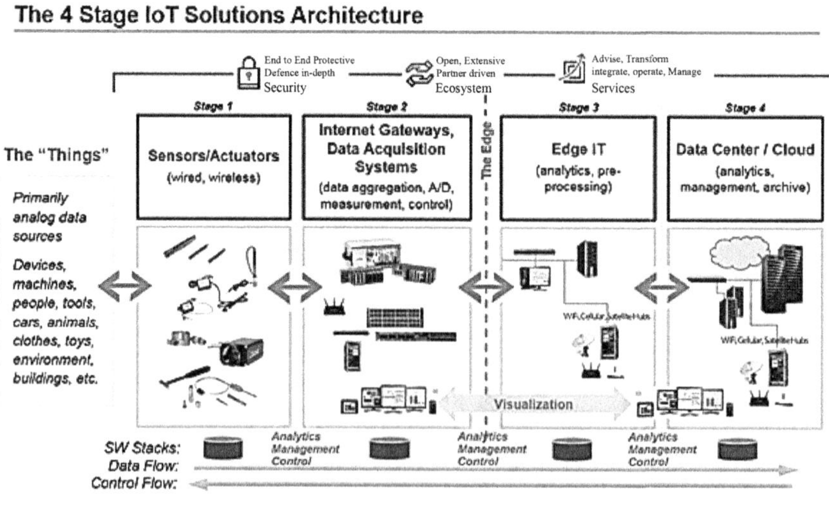

FIGURE 7.6 IoT energy domain.

7.5.1 RESIDENTIAL ENERGY

As technology advances, the cost of energy increases as well. Consumers seek ways to cut and control their energy bills. IoT enables a mature method of analyzing and optimizing the use of a gadget and the overall system of a household. It could be as simple as adjusting the device's setting, turning it on and off, or reducing lights to conserve energy [12].

Energy waste has a significant impact on the price of production of every business operation. IoT enables a unique method of monitoring and treatment while maintaining a cheap cost and a high grade of care. IoT enables an acceptable method of monitoring energy costs and boosting organizational performance. It identifies energy issues as functional issues inside a complex business network and proposes solutions [13–15].

7.5.2 TRUSTWORTHINESS

By delivering data and action, the IoT feature provides system stability. It identifies hazards to the device's efficiency and dependability, thereby protecting against losses caused by malfunctioning devices, outages, and damage.

7.5.3 SMART OBJECTS IN IOT

The perception of elegance in IoT is worn for the physical substance that is active, digital, and networked, can maneuver to various quantities in competition, reconfigurable, and has confined the possessions. The chic stuff needs energy, data storage, etc. [16,17]

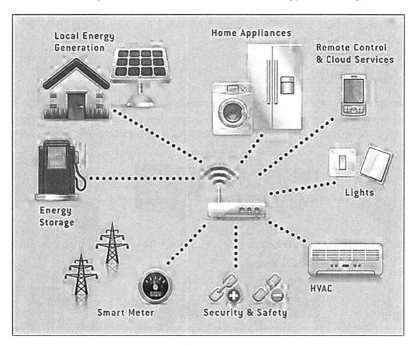

FIGURE 7.7 Commercial energy.

A smart object is an entity that augments the interface with elegant supplementary stuff as well as with people also. IoT is the association of consistent assorted objects (such as smart devices, smart objects, sensors, actuators, RFID, and embedded computers) distinctively addressable and foundation on standard announcement protocols (Figure 7.7).

In a chronic life, people boast a bundle of the object with the Internet or wireless or wired connection. Such as

* Smartphone,
* Tablets, and
* TV computer

These substances can be consistent among them and assist our daily life (smart home, smart cities) no fabric the circumstances, localization, convenience to a sensor, size, scenario, or the risk of danger.

Smart objects are exploited extensively to convert the substantial milieu just about us to a digital world using the IoT technologies. An elegant article clutch hunk of relevance reason constructs sagacity for their limited circumstances and intermingle with human users. A chic object sense, log, construe the incidence within themselves and the milieu, and intercommunicate with apiece further and replace the information with people.

The exertion of chic objective has alerted the procedural features (such as software infrastructure and hardware platforms) and claim scenarios. Application areas were sorted from supply-chain administration and endeavor applications (home and hospital) to healthcare and engineering agencies. As for human boundary characteristics of smart-object technologies are immediately commencement to collect consideration from the atmosphere.

7.6 IOT DEVICES

IoT devices are an atypical campaign that bonds wirelessly to a complex with a piece supplementary and intelligent to convey the data. IoT diplomacy has broadened the Internet connectivity afar typical diplomacies such as smartphones, laptops, tablets,

Data Collection Devices   Smart Machinery   Phones and Tablets   Home Automation

RFID Systems   Digital Signage   Security Systems   Medical Devices

FIGURE 7.8 Various smart objects.

and desktops (Figure 7.8). Implanting these devices with technology facilitates us to commune and cooperate over the networks, and they can be distantly scrutinized and prohibited.

There is a hefty diversity of IoT devices accessible foundation on IEEE 802.15.4 standard. These campaign assortments from wireless motes, separable sensor boards to interface board, are practical for researchers and developers.

IoT strategy embraces computer devices, software, wireless sensors, and actuators. These IoT strategies are allied over the Internet and facilitate the data relocation accompanied by objects or people robotically without human intervention [18].

7.7 PROTOCOLS ON THE IOT NETWORK LAYER

The edifice stratum is estranged into two sublayers: A routing layer that separates the packet transmission from spring to target and an encapsulation layer that defines the packets (Figure 7.9).

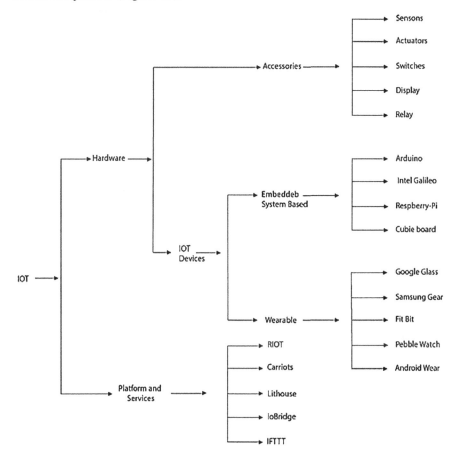

IoT Devices and Technologies

FIGURE 7.9 Communication protocol.

7.7.1 PROTOCOL TO THE RPL

RPL is an abbreviation for Low-Power Latitudinal Routing Protocol with Lossy Network. This is a distance-vector exercise designed to simplify the rules imposed by Data Link wires. RPL generates a Destination-Oriented Guided Acyclic Graph (DODAG) with a single path to the origin originating with each leaf bump. All traffic is routed through the DODAG's root. Each node initially transmits a DODAG Information Object (DIO) identifying as its root. These are steadily increased through network twist actions and absolute DODAG. Assume that a pioneering knot obstructs the network's induction. In that scenario, it issues a request for DODAG Knowledge Solicitation (DIS) and springs counter vertebral with a DAO Acknowledgement (DAO-ACK) to verify the join.

7.7.2 CORPL PROTOCOL

The CORPL procedure is the RPL protocol leeway, which is marked as cognitive RPL. The DODAG topology influences this complicated prescription for observant inference and use. The technique of CORPL develops two groundbreaking variations in the practice of RPL. Amid the nodules, it munches malleable front to a box ahead. Just as parents retain it, each bump of the CORPL set of rules lingers slightly set in frontward categorization. Nodule refashions its adjustment to its universal use of DIO texts, respectively. Every node typically modernizes the general situation aimed at the invariable set forwarder at the root of this restructured communication.

7.7.3 PROTOCOL CARP

A Scattered Routing Protocol is a CARP (Channel-Aware Routing Protocol). Sunken contact is cautious. It has a little kit to be used for the IoT. It accomplishes two unusual functions: network initialization and front-end data. CARP protocols do not hold data composed beforehand. Therefore, for specific IoT or other applications where data are updated regularly, it is not helpful. The CARP upgrade is carried out in E-CARP, which triumphs over the CARP restriction. The E-CARP permits the descending node to set aside formerly conventional sensory data.

7.7.4 6LoWPAN

The mentioned protocol of 6LoWPAN works on IPV6 with Low-Power Personal Area Network, which has the capacity of transmitting over low data rate network applied with lightweight IP-based communication. There might be insufficient dispensing facilities to transmitting the proof using an Internet protocol wirelessly. So, for home and building automation, it is primarily used. They use nearly 250 kps transfer rate, in this protocol 6LoWPAN operates within the 2.4 GHz frequency range. The maximum bit packets will be 128-bit packets in the header that were used as maximum.

7.7.5 6LoWPAN Protection Measure Calculation

For the 6LoWPAN contact protocol, Haven is a primary concern. There is a copious attack disorder at the safekeeping stage of 6LoWPAN that aims to demolish the network continuously. Since it combines due to the existence of two systems, there is a risk of assault from two directions that targets the entire 6LoWPAN layer stack (Physical layer, Data Link layer, Adaptation layer, Network layer, Transport layer, and Application layer).

7.7.5.1 Properties of the Protocol of 6LoWPAN

- **Standard**: RFC6282
- **Frequency**: worn beyond a multiplicity of further network medium counting Bluetooth Smart (2.4 GHz) or Zigbee or low-power RF (sub-1 GHz)
- **Range**: NA
- **Data rates**: NA

7.8 PROTOCOLS ON THE IOT SESSION LAYER

The session layer protocols examine the message passing needs and procedures. Numerous standardization bodies have proposed protocols for the IoT session layer. There are numerous session layer protocols available, each with a unique set of capabilities and capabilities. MQTT and CoAP meet these requirements through short message sizes, message management, and low message cost [20].

7.8.1 MQTT (Message Queue Telemetry Transport)

MQTT is a messaging protocol that was inaugurated by IBM in 1999. It remained predominantly rigid for examining projection nodes besides quarantined chasing in IoT. Its outfits are insignificant, inexpensive devices with low memory and low power. MQTT bequeath with surrounded Between Connectivity relevance and middleware in an isolated area and alternative crosswise it bonds Communicators and Networks [21].

The MQTT protocol is based on the architecture of publish/subscribe. There are three critical components of the publish/subscribe architecture: authors, subscribers, and a broker. Bestowing to the IoT pinnacle of examination, originators are frivolous sensor expedients that lob, whenever practicable, their data to a concomitant advisor and serves rump to sleep. Subscribers are apps that are intrigued by a specific subject or sensory data, so when new data are obtained, they are amalgamated into brokers to be conversant. The broker acknowledges the sensory data, riddles them on particular subjects, and sends them to subscribers depending on their interest in the subjects.

7.8.2 SMQTT

Secure Message Queue Telemetry Transport (SMQTT) is a conservatoire of MQTT protocol whichever achievement encryption supports inconsequential representative encryption. The significant enhancement of this encryption is because it has

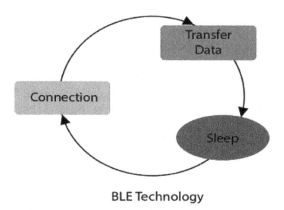

BLE Technology

FIGURE 7.10   SMQTT (Secure Message Queue Telemetry Transport).

a function of simulcast encryption. Under this functionality, one memorandum is encrypted to many other nodes and delivered. The development of memorandum conveys and being paid (Figure 7.10). Four main phases are as follows:

1. **Setup**: During this point, the commissioners and subscribers' inventory to the negotiator himself and get a surreptitious master key.
2. **Encryption**: Once the evidence is circulated to the adviser, it is expressed in code by the negotiator.
3. **Publish**: The broker is responsible for distributing the encrypted file to the subscribers.
4. **Decryption**: Lastly, the time-honored memorandum subscribers with an identical master key are decrypted.

MQTT is proposed only to boost the security function of MQTT.

7.8.3   CoAP

Constrained Application Protocol (CoAP) is a session layer performance which mark offered the RESTful (HTTP) boundary amid HTTP client and server. It is deliberate through IETF mortified RESTful Environment (CoRE) practical assemblage. It is deliberated to custom campaign on the matching discomfited network among devices and wide-ranging nodes over the Internet. CoAP expedites low-power sensors to consume RESTful services, whereas discussing their stumpy influences constrictions. The etiquette is explicitly erected on behalf of IoT systems principally pedestal on HTTP protocols [22].

Such a network is worn encircled by regional networks or hips discomfited environs. CoAP's entire schedule consists of the CoAP client, CoAP server, REST CoAP proxy, and Internet REST.

The information is revealed to the CoAP server from CoAP clients (such as smartphones and RFID sensors), and the comparable memorandum is withdrawn to the REST CoAP proxy. The REST CoAP proxy correlates the condition of the CoAP externally and uploads the information through the REST Internet (Figure 7.11).

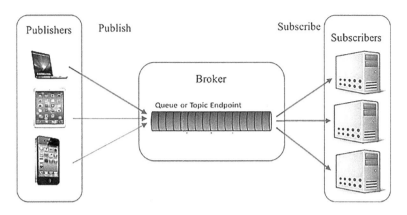

FIGURE 7.11   CoAP (Constrained Application Protocol).

7.8.4   DATA DISTRIBUTION SERVICE (DDS)

Data Delivery Service (DDS) is a communication protocol middleware (sometimes called machine-to-machine (M2M)). The Object Management Group (OMG) usually accomplishes this with high-speed and high-performance, scalable, secure, and interoperable data sharing for real-time coordination. The practice of this communiqué is based on a publish-subscribe pattern for transmission and payment between nodes of knowledge, events, and commands.

The DDS etiquette has two main layers:

The Data Distribution Service (DDSTM) is an Object Management Group (OMG) infrastructure and API standard for data-centric networking.

a. **Data-centric publish-subscribe (DCPS)**: This stratum distributes the in sequence to subscribers.
b. **Data-local reconstruction layer (DLRL)**: This deposit affords a border to DCPS functionalities, sanction such allotment to scattered facts amid IoT permit things.

7.9   JAYA OPTIMIZATION

The Jaya algorithm is a clone that solves both close and unconstrained challenges with optimization. It is a cultural approach that varies a range of specific solutions repeatedly. It is an optimization method with no gradients. There are various algorithms for optimization such as genetic algorithms, the optimization of the particle swarm algorithms, artificial colony optimization algorithms, firefly optimization algorithms, biogeographic optimization algorithms, and cuckoo search. (Figure 7.12).

Jaya is a globally recognized simple and robust optimization [15] framework with application to the benchmark role of both unconstraint and constrained problems. Though it is parameterless like the TLBO, it differed from TLBO by not requiring a learner phase, whereas TLBO uses both learner and teacher phases [16]. Jaya is based on achieving a given problem by sidestepping the foulest elucidation and moving toward the superlative elucidation. This algorithm is excellent as it requires

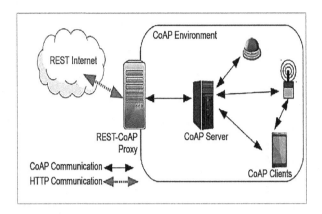

FIGURE 7.12   Flowchart of proposed algorithm.

only a few parameters of regulation, like the number of design variables and the overall population size/generational number. There is no need for any specific algorithmic control parameter and does not need extensive parametric tuning before executing computational experiments. The framework of Jaya is charted in an easy-to-understand manner in Algorithm-1. Tenancy $f(x)$ represents the objective function (OF) that needs to be lessened. Assume the number of design variables to be "n" ($j=1$ to n) and "m" to be the number of individual solutions ($k=1$ to m) at any given iteration i. Furthermore, let $f(x)$ be the candidate who obtains the best solution in a population, while the nastiest candidate achieves the nastiest of values $f(x)$. If for ith iteration Xj, k, i is the jth adjustable value for the kth candidate, this value Xj, k, i is altered using Equation 7.1. X's, k, $i = Xj$, k, $i + r1$, j, $i(Xj$, best, i-$|Xj$, k, $i|$)$-r2$,j, $i(Xj$, worst, i-$|Xj$, k, $i|$)... (1) where, $X'j$, k, I = the updated new Xj, k, I value, best, I = best Xj, k, I value, Xj, worst, i = worst Xj, k, I value, $r1$, j, i and $r2$, j, i = two random numbers ranging from 0 to 1. The value of X's, k, i is considered acceptable if it offers a healthier OF, which will serve as the subsequent iteration input. This cycle is continuously repeated until the algorithm is stopped after reaching a predetermined number of iteration [23–26].

7.9.1 PROPOSED ALGORITHM

The steps of the proposed algorithm are as follows:
    Input:

1. Sample data Xi from data streaming without label or target (unsupervised learning).
2. Use the Jaya algorithm to obtain the optimal Dr value. Output: Shortest path results in C.

Method 1: Apply the Jaya algorithm to select the best Dthr value of ECM according to the input data.

    Method 2: Apply the Jaya algorithm on the shortest path parameters.

    Method 3: Establish the first cluster C1 by taking X1 as the first cluster center (Cc1) with a value of initial cluster radius (Ru1) value of 0 as the first data point entered.

Method 4: If all the data samples in the data stream have been disposed of, the method is terminated. Else the distance between the current data entered xi, and all the existing cluster centers Ccj is calculated using the formula $Dij = \|xi - Ccj\|$, $j = 1$, $2,..., k$; where k represents the number of values in prevailing clusters.

Method 5: If $Dij.Ruj, j = 1, 2,..., k$; this means that the current Dim Ccj data input) xi belongs to the existing Cm cluster, -, $k(xi$ min1–, 2, j Ccm and in this case, xi Cm. If no new cluster is formed and no existing center or radius of the cluster is changed, go back to step 2; otherwise, proceed to the next step.

Method 6: Regulate the Dij amount of the current input data xi and all existing cluster centers Ccj and the Ruj radius of the corresponding cluster, i.e., $Sij = Dij + Ruj$, $j = 1, 2,..., k$. Select a Ca cluster center (Cca is the appropriate cluster center and Rua is the radius) and make it comply with $Sia = Dia + Rua = \min(Sij)$, $j = 1, 2,..., k$.

Method 7: If Sia > 2×Dr, xi is not part of any existing cluster. To create a new cluster, use Step 1, and then, go back to Step 2. Step 8: If Sia-2-Dr, removing its Cca and adding its Rua, xi-Ca, and Ca are modified. The new radius of the cluster Runewa = Sia/2; subtract the new center of the cluster Ccnewa from the xi and Cca ligature to satisfy the condition $\|Ccnewa - xi\| = Runewa$; then, return to step 2.

7.9.2   PROPOSED MODEL DIAGRAM

The proposed channel assignment has three processes (Figure 7.13).

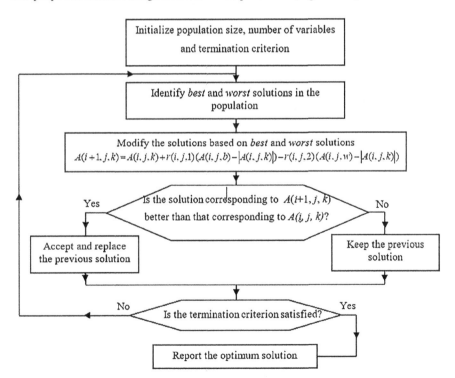

FIGURE 7.13   Proposed model diagram.

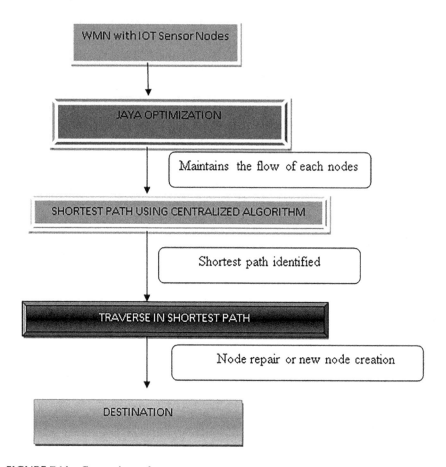

FIGURE 7.14   Comparison of energy.

The comparison of Energy is shown in figure 7.14.

7.10   RESULT ANALYSIS

The WMN supports a large number of nodes, but for the implementation of testing, we compare 20, 40, 60, 80, and 100 nodes at different times at $2500 \times 2500\,\mathrm{m}$ (Table 7.1). In Ns-2 ver. 2.31, we chose to conduct the simulation. While the dominant network simulation tool has been Network Simulator 2 (NS2), it does not have native multichannel simulation support. The two values on TCL and C++ codes, modifications are carried out. Ns-mobile node. Tcl and ns-lib. The TCL level is updated to assemble multichannel components and make TCL scripts to sustenance multichannel arrangement. To add the cross-layer-dependent channel selection algorithm and control the multichannel node lists, MAC and physical layer relevant files are updated at the C++ level [26–29].

Energy level determines the lifetime of a node in the network. When compared with EMTX and NGA, the JAYA shows an 85% increase with EMTX and a 77% increase with NGA.

TABLE 7.1
Simulation Parameters

Parameter Type	Values in Simulation
Total time for simulation	200
Range of nodes	Varied between 20 and 100
Method of traffic	Constant bit rate
Pause time	10 seconds
Topology area	2500×2500
Size of packets	512 bytes
MAC protocol	IEEE 802.11
Mobility model	Random waypoint
Wireless channel	Physical wireless
Type of antenna	Omnidirectional
Queue type	Drop tail

7.11 CONCLUSION

The proposed technique shows higher stability of energy comparison when compared with previous techniques. The Jaya optimization algorithm can be slightly modified and updated for QoS parameters such as throughput, error rate, jitter, and scalability.

REFERENCES

[1] Revathy, G., Parimalam, A., and Kanishka, I. (2020) "Diabetic detection using Irish", *Int J Sci Res Eng Manage (IJSREM)* 4(3), 2582–3930.

[2] Revathy, G., Kavitha, N. S., Senthivadivu, K., Sathya, D., and Logeshwari, P. (2020) "Girl child safety using IoT sensors and tabu search optimization", *Int J Recent Technol Eng (IJRTE)* 8, 2277–3878.

[3] Revathy, G., Saravanan, G., Arieth, R. M., and Vengateshwaran, M. (2019) "Magnify Qos with tabu & link scheduling in WMN", *Int J Recent Technol Eng (IJRTE)* 8(4), 2277–3878.

[4] Revathy, G. (2018) "Mounting eminence of services in wireless mesh networks", *Int J Res Anal Rev* 5(3), 43–51.

[5] Revathy, G., and Selvakumar, K. (2018) "Sustain route by tabu and amplified QoS by distributed scheduling in WMN", *Int J Recent Trends Eng Res* 8(1), 7–12.

[6] Revathy, G., and Selvakumar, K. (2018) "Channel assignment using tabu search in wireless mesh networks", *Wireless Personal Commun* 100(4), 1633–1644.

[7] Revathy, G., and Selvakumar, K. (2018) "Increasing quality of services in wireless mesh networks", *Int J Adv Res Comput Eng Technol* 7(3).

[8] Revathy, G., and Selvakumar, K. (2018) "Escalating quality of services with channel assignment and traffic scheduling in wireless mesh networks", *Cluster Comput* 25(5), 11949–11955.

[9] Revathy, G., and Selvakumar, K. (2017) "Route maintenance using tabu search and priority scheduling in wireless mesh networks", *J Adv Res Dyn Control Syst* 9(6).

[10] Holm, S. (1979) "A simple sequentially rejective multiple test procedure", *Scand J Stat* 6(2), 65–70.

[11] Joaquin, D., Salvador, G., Daniel, M., and Francisco, H. (2011) "A practical tutorial on the use of nonparametric statistical tests as a methodology for comparing evolutionary and swarm intelligence algorithms", *Swarm Evol Comput* 1(1), 3–18.

[12] Karaboga D., Basturk B. (2007) Artificial Bee Colony (ABC) Optimization Algorithm for Solving Constrained Optimization Problems. In: Melin P., Castillo O., Aguilar L.T., Kacprzyk J., Pedrycz W. (eds) Foundations of Fuzzy Logic and Soft Computing. IFSA 2007. Lecture Notes in Computer Science, vol 4529. Springer, Berlin, Heidelberg. https://doi.org/10.1007/978-3-540-72950-1_77

[13] Kumari, N. V., Ghantasala, G. S. P., and Arvindhan, M. (2020) "Compulsion for Cyber Intelligence for Rail Analytics in IoRNT", *Securing IoT and Big Data.*

[14] Karaboga, D., and Akay, B. (2009) "A comparative study of artificial bee colony algorithm", *Appl Math Comput* 214(1), 108–132.

[15] Arvindhan, M., and Anand, A. (2019) "Scheming an proficient auto scaling technique for minimizing response time in load balancing on Amazon AWS cloud", *SSRN Electron J* doi: 10.2139/ssrn.339080.

[16] Liang, J. J., Runarsson, T. P., Mezura-Montes, E., Clerc, M., Suganthan, P. N., Coello, C. A. C., and Deb, K. (2006) Problem Definitions and Evaluation Criteria for the CEC 2006 Special Session On Constrained Real-Parameter Optimization, Technical Report, Nanyang Technological University.

[17] Singapore, http://www.ntu.edu.sg/home/EPNSugan.

[18] Patel, V. K., and Savsani, V. J. (2015) "Heat transfer search (HTS): a novel optimization algorithm", *Inf Sci* 324, 217–246.

[19] Rao, R. V. (2015) *Teaching Learning Based Optimization and Its Engineering Applications.* London: Springer Verlag.

[20] Muthusamy, A., Anand, A., Kannan, T. V., and Rao, D. N. (2021) "The modern way for virtual machine placement and scalable technique for reduction of carbon in green combined cloud datacenter", *Green Energy Technol.*

[21] Rao, R.V., and Patel, V. (2013) "Comparative performance of an elitist teaching-learning-based optimization algorithm for solving unconstrained optimization problems", *Int J Ind Eng Comput* 4(1), 29–50.

[22] Rao, R. V., Savsani, V. J., and Vakharia, D. P. (2011) "Teaching-learning-based optimization: a novel method for constrained mechanical design optimization problems", *Comput-Aided Des* 43(3), 303–315.

[23] Pandey, H. M., Chaudhary, A., and Mehrotra, D. (2014) "A comparative review of approaches to prevent premature convergence in GA", *Appl Soft Comput* 24, 1047–1077.

[24] Pandey, H. M., Chaudhary, A., and Mehrotra, D. (2016) "Grammar induction using bit masking oriented genetic algorithm and comparative analysis", *Appl Soft Comput* 38, 453–468.

[25] Rao, R. (2016) "Jaya: a simple and new optimization algorithm for solving constrained and unconstrained optimization problems", *Int J Ind Eng Comput* 7(1), 19–34.

[26] Anand, A., and Arvindhan, M. (2020) Development and various critical testing operational frameworks in data acquisition for cyber forensics, In *Critical Concepts, Standards, and Techniques in Cyber Forensics*, IGI Global, pp. 88–102.

[27] Shikha, S. (2020) "A comparative analysis for prediction of Parkinson's diseases using classification algorithm", *Int J Res Appl Sci Eng Technol* 8(7). doi: 10.22214/ijraset.2020.7057.

[28] Arvindhan, M., and Ande, B. P. (2020) "Data mining approach and security over ddos attacks", *ICTACT J Soft Comput* 10(2). doi: 10.21917/ijsc.2020.0292.

[29] Anand, A., Chaudhary, A., and Arvindhan, M. (2021) "The need for virtualization: when and why virtualization took over physical servers", *Lecture Notes Electr Eng* 668. doi: 10.1007/978-981-15-5341-7_102.

8 Data Security Essentials for the Convergence of Blockchain, AI, and IoT

Siva Shankar Ramasamy
International College of Digital Innovation -
Chiang Mai University

S. Vijayalakshmi
CHRIST (Deemed to be University)

S. P. Gayathri
The Gandhigram Rural Institute (Deemed to be University)

Nopasit Chakpitak
International College of Digital Innovation -
Chiang Mai University

CONTENTS

DOI: 10.1201/9781003081180-8

8.1   INTRODUCTION

Significant innovations of computerized revolutions are blockchain, Internet of things (IoT), and artificial intelligence (AI). These technological advances are perceived as developments that can improve current business measures and make new plans of action. Blockchain, for instance, can expand reliability, straightforwardness, security, and protection of business measures by giving a mutual and decentralized disseminated record. In general, a blockchain is, or for the most part, a disseminated record that can store a wide range of resources like a register [1]. Essentially, this information can be identified with cash and personalities. IoT steers the mechanization of enterprises and ease of use of industry measures that are fundamental for international businesses. At long last, AI gets better measures by identifying designs and enhancing the results of these trade measures [2]. As yet, the interconnection between these three developments is frequently dismissed, and blockchain, IoT, and AI are ordinarily utilized independently. Be that as it may, these developments can and ought to be applied together and will unite later on. A potential association between these advances could be that IoT gathers and gives information, blockchain gives the framework and fixes the guidelines of commitment, while AI streamlines practices and tenets [2,3]. By plan, the above-said advancements are corresponding and can abuse their maximum capacity whenever joined.

Bringing together at one common place for preparing information and dynamic are moving toward decentralization to improve security. New specialists are likewise arising, forming a short term set apart by the insight of machines and the proficiency of foundations. At the point when these three advances are dissected independently, it tends to be seen how AI is changing workplaces, forcing effectiveness on arranging and performing profitable undertakings in a wide range of mechanical exercises with the summed-up combination of robots. Specifically, blockchain is decentralizing [4] the monetary and budgetary areas and showing its potential for application in different regions, though IoT is prevailing with regard to changing the Web into an instrument going about as a transmission belt in the execution of examination-based AI measures.

8.2   THE PREAMBLE OF BLOCKCHAIN

Blockchain is also known as distributed ledger technology (DLT) [5], which makes the historical backdrop of any computerized resource unalterable and straightforward using decentralization and cryptographic hashing. For example, when we use Google document, it makes an archive and offers it with a gathering of individuals, and the report is circulated rather than replicated or moved. This makes a decentralized dissemination chain that provides everybody admittance to the record simultaneously. Nobody is bolted out anticipating changes from another gathering, while all alterations to the documents are being witnessed continuously and such modifications are apparent (Figure 8.1).

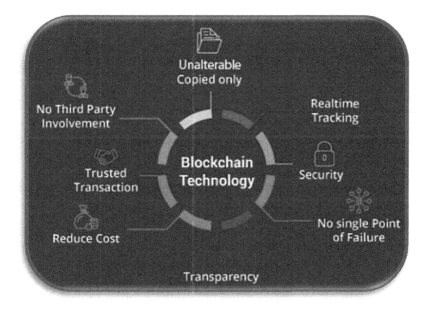

FIGURE 8.1   Advantages of blockchain.

8.2.1 ADVANTAGES OF BLOCKCHAIN

8.2.1.1 Transparency

Exchange accounts are getting more straightforward using blockchain innovation. Since blockchain is a kind of conveyed record, all organization members share similar documentation instead of individual duplicates. That mutual form must be refreshed through agreement, which implies everybody must concede to it. To change a solitary exchange record would require the adjustment of every resulting record and the conspiracy of the whole organization. In this manner, information on a blockchain is more precise, predictable, and straightforward than when it is pushed through a paper-based formal procedure. It is additionally accessible to all members who have permission to access it.

8.2.1.2 Security

Blockchain technology offers various ways to make records safer than other similar frameworks. In blockchain, exchanges of data must be settled, before they have been saved. Later, the exchange of data is affirmed, and it is scrambled and connected to the past exchange. It is in conjunction with the way that data are put away over an organization of PCs rather than on a common single computer system, which creates extremely hard for programmers who hack information on the Internet. In some companies, securing confidential information is difficult. In such a situation, blockchain provides chances to secure significant data from the fraudulent movement.

8.2.1.3 Tracking

Tracking information is one of the advantages of blockchain. At the point when trades of products are stored on a blockchain, it can be reviewed to confirm where a resource arrived from and information about each halt as well. This chronicled exchange of information can assist with checking the legitimacy of resources and forestall scams.

8.2.1.4 Reduce cost

In general, any organization prefers cost-benefit by lessening expenditure. Using blockchain technology, there is no requirement for the same number of outsiders or mediators who assure that it does not make a difference in the event for which any businessman can believe his alliances. All things considered, we can simply need to confide in the information on the blockchain. Additionally, there is no need for auditing such a huge amount of documentation to finish an exchange since everybody will have consent admittance to a solitary, changeless variant.

8.2.1.5 Effectiveness and Speed

In conventional business, exchanging no matter which we want is a tedious cycle that is inclined to a manual fault and regularly requires outsider intervention. To make the process of trading with blockchain will be automatization. These exchanges can be finished more proficiently and quicker. Even though the information is recorded with a single ledger, it will be distributed between members. It is not necessary to accommodate numerous records and can be finished with not as much of a mess. What's more, when everybody approaches similar data, it gets simpler to confide in one another devoid of the requirement for various mediators. Along these lines, sorting out things can happen a lot snappier.

8.3 THE PREAMBLE OF ARTIFICIAL INTELLIGENCE (AI)

AI [6] is the broad range of software engineering for developing machines equipped for achieving jobs that normally need individual intellectual. AI is an interdisciplinary science with different methodologies; however, innovations in machine learning and deep learning are making a change basically in each area of the information technology field.

8.3.1 Various Ways of Perception on Artificial Intelligence

   I. AI collects information from their general surroundings and then implements such information in various areas. For instance, neural network-based video games prove that the way of playing games by computer is the same as people.
  II. Natural language processing (NLP) permits machines to peruse, comprehend, and decipher individual languages. It utilizes all measurable techniques along with semantic programming to know grammar and syntax.
 III. A recent couple of years, the devices that have sensors such as video cameras, microphones, device locating GPS, and location finder radar system are having been observed like a machine, which also includes voice recognition and computer graphical vision to recognize various things.
  IV. Machine-oriented robotic gadgets are generally utilized in large industries for production, multi-specialty medical centers, etc. In recent times, unmanned aircraft and drones have depended on a high degree of complexity to explore the various processes by utilizing machine view.
   V. Digital assistance is providing amazing support in the systems like automatic driving cars and robots, and computerized collaborators, for example, Alexa and Siri, are popular digital assistants in smart devices and they need harmony to understand the activities of people with accepted communal practices.

The key benefits of AI are as follows:

- AI reduces time to complete a job. It empowers performing various tasks at a time and facilitates the outstanding burden for existing assets.
- AI allows the execution of tasks continuously without disruption.
- AI enhances the capacities of diversely able people.
- AI has a collection of promotions across business industries.
- AI makes the process easy and makes the cycle quicker in taking decision (Figure 8.2).

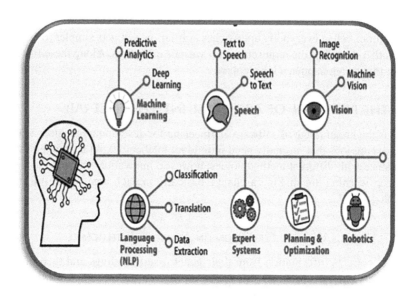

FIGURE 8.2 Artificial intelligence.

8.4 THE PREAMBLE OF THE INTERNET OF THINGS (IOT)

IoT is an environment of connected devices items that are open through the Web. The "thing" in IoT [7] might be an automated device with sensors embedded in it, for example, the thing that has been allotted an IP address and can gather and move information over the Internet without manual help or mediation. This built-in innovative technology encourages them to connect with the interior position or the outer atmosphere, which thus influences the choices taken. Few benefits of IoT are as follows (Figure 8.3):

FIGURE 8.3 Internet of things (IoT).

8.4.1 Benefits of the Internet of Things (IoT)

8.4.1.1 Lessen in Cost

Many companies use IoT gadgets to smooth out activities and increment benefits. These additional benefits of the IoT innovations will be custom-made to enable such companies to accomplish something. From Internet safety to working environment effectiveness, IoT gadgets are as of now affecting organizations' primary concerns. Supporting expenses can be decidedly affected than IoT gadgets, which are having sensors to maintain company equipment and moving toward top proficiency, while progressing investigating company hardware gets issues before influencing working members and sparing the problem and expenses of huge fixes. This limits exorbitant broadened personal time for fixing the issues.

IoT is amazingly helpful to organizations in production, transportation, assembling, coordination, etc. Various approaches to utilize IoT innovation are available, which will be well effective to the primary concern of an organization through smoothing out regular working cycles, and it will be the peak teamster of IoT venture in some organizations (Figure 8.4).

8.4.1.2 Efficiency

At any workplace, IoT can be utilized to upgrade a company's work plan. The office assets can be implemented with IoT technology and placed to be used at any location within the premises for a superior work process.

8.4.1.3 Company Prospects

Numerous organizations endeavor to get to the income-creating intensity of computerized administrations, but the majority do not have a strong procedure for seeking after this road. IoT provides a distinct advantage in this regard, as cutting-edge examination, AI, and shrewd utility networks construct this work as simple for small

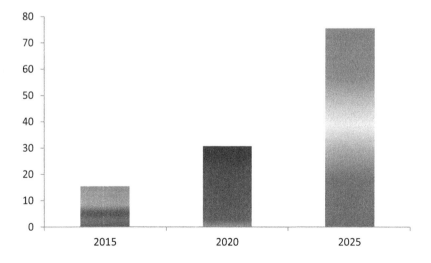

FIGURE 8.4 Number of IoT devices around the world between 2015 and 2025 [8].

to medium businesses to gather significant information, which is expected to offer the benefits for their clients who are looking for. These examinations give innovative experiences, which are not exclusively being used to make novel plans of action, however totally reclassifying the conventional businesses.

8.5 CONVERGENCE OF BLOCKCHAIN AND AI

Blockchain and AI [9] both have a place to establish of novel business age. Even though blockchain is demonstrated to be the best innovation in the commercial center to put resources into, it additionally has a few impediments. For this reason, AI will give the correct answer to change blockchain, an ideal advancement. When AI and blockchain converge, both will settle different difficulties and think of a heap of fresh chances (Figure 8.5).

8.5.1 Benefits of Blockchain When Coupled with AI

8.5.1.1 Enhanced Decision-Making

The collaborative development of AI and blockchain innovation will advance decision-making capacity. AI can identify the fakes and dangers related to any exchange or application, and blockchain can help to increase the exactness of the information utilized for fake identification associated with AI procedures, which have basic and safe information inspecting methods.

8.5.1.2 Protected Customized Understanding

A combination of AI and blockchain could develop the customized understanding, and it will be conveyed to the clients. At the same time, AI will keep on progress with profoundly delicate information of everyone; in the case of blockchain, it will guarantee that the information stays unchanging and secures utilizing encrypted methodology. This feature will help to secure the conditions, while AI neglects to provide the privileges of customized understanding for the clients, because of information infringement made by the gate-crashers.

FIGURE 8.5   Blockchain and AI.

8.5.1.3 An Extreme Level of Trust

Consolidating AI and blockchain advancements will engage clients to investigate the procedure in which their secret information is saved, maintained not with any negotiations on the protection and security level. This feature will assist the clients to review every progression in the decision-making method.

8.5.1.4 Expanded Accessibility

These two technologies jointly will reduce the problems related to installment techniques. Blockchain-related digital forms of money will set the cash obstruction to finish and cultivate global exchanging; meanwhile, AI will progress the functional proficiency of procedure, guarantee advanced security, and lessen the expense.

8.5.1.5 Innovative Business Strategy

The prominent advantage of these two innovations is the presentation of more up-to-date plans of business action. The data of the total business environment, including entire business partners, are supplied by blockchain, for providing permanent information without agonizing over the responsibility for the system. This feature will furnish a constructive AI framework including further bits of knowledge from the examples, practices, and different variables identified with the functioning of a successful and competitive business. Also, accordingly, it conveys with all the more verifiably exact choices, in other words, more up-to-date plans of action.

8.5.1.6 Enhanced Smart Business Contracts

Smart business contracts are presenting different advantages such as top speed, least to zero disagreements, and better storage for information to a smart business empire. However, such smart contract's utilization is limited because of the complexity of the program. When computer-based intelligence like AI is incorporating with blockchain, this innovative partnership will make business complexity, feasible by secret coding, and then approve complex business connections on a blockchain. Hence, enhanced smart business contracts will be possible and appear.

8.6 CONVERGENCE OF AI AND IOT

The primary reason for incorporating AI with IoT [10] is to accelerate profitability. It is necessary for delivering excellent merchandise at a quick rate. Furthermore, ventures require to deliver imaginative plans as well. In such a situation, there is a strain to decrease the creation cost even with the entire limitations. It results in expanding complexity nature of the hardware utilized. One explanation for these problems is AI. It can build the proficiency of the business and execute computerization (Figure 8.6).

A few organizations have just embraced these two technologies as a component of their business cycles and final products. It likewise established this pair of technology for companies where they wish to invest to reach the pinnacle of innovations by implementing and adopt a competitive benefit. Figure 8.7 shows recent innovative technologies adopted by industries.

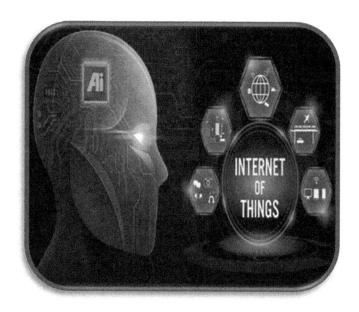

FIGURE 8.6   AI and IoT.

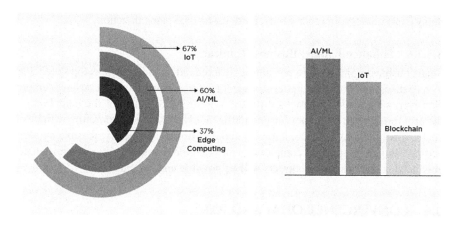

FIGURE 8.7   State-of-the-art technologies exercised by industry.

8.6.1   Fruitful Applications of AI and IoT in Industries

8.6.1.1   Robot Support for Industry Production

Manufacturing and production are significant processes in all industries. If they adopted new advanced technologies such as AI and IoT, then robots will be more convenient for their business processes. Robots are having embedded with sensors, which encourages information communication. In addition, when the robots are implemented with AI for reasoning and calculations, they can gain more up-to-date information. This methodology spares time and expenditure. This results in manufacturing and production on time.

8.6.1.2 Self-Supporting Automated Vehicles

Self-supporting vehicles are the best illustration of IoT and AI cooperation. The strong knowledge of AI helps such vehicles to anticipate the conduct of people when they walk on a road with different conditions. For instance, smart vehicles can decide street rules, ideal vehicle speed, and climate, thus accomplishing more intelligence during travel.

8.6.1.3 Investigating Retail Information

The retail investigation includes various information from digital cameras and image sensors to notice clients' mobility, and later, this system will identify when the clients will arrive at their destination.

8.6.1.4 Smart Temperature

The incorporation of mobile communication devices can verify and deal with the temperature from any place dependent on the job routine and temperature inclinations of its clients. Hence, the combination of AI with IoT innovation can show the way of a new degree of clarification and experience.

8.7 CONVERGENCE OF IOT AND BLOCKCHAIN

The field of science and innovation has consistently advanced and provides numerous benefits. This advancement provides better answers for people who are searching for solutions for their issues. The most recent couple of years have been the same with the headways in innovative fields such as AI, IoT, and blockchain. Among all these technological development, the convergence of IoT and blockchain [10] has shown potential support.

IoT has a great deal of effective executions, and numerous areas are being affected by IoT innovation, such as medical care, education, agriculture, and even in smart homes. Significantly, IoT concerns security and protection, and for this reason, this innovative technology has arrived at another degree of status lately. From that point forward, no critical improvements have been accounted for (Figure 8.8).

FIGURE 8.8 IoT and blockchain.

The other technological development, blockchain, can change advanced technological revolutions in organizations everywhere, demonstrating constructive outcomes. However, the reality requires proving that blockchain should yet be tried on genuine tasks. Hence, financial specialists are having still an issue of whether they will receive a legitimate rate of profitability or not. Albeit these two advancements are fruitful in their way, converging IoT and blockchain will make the way for a fate of boundless potential benefits.

8.7.1 Benefits of IoT-Enabled Blockchain

8.7.1.1 Security

One of the significant obstructions of IoT execution is security. Presently, IoT could utilize support from blockchain and also its vigorous encryption guidelines. This will acquire a more grounded level of security to the innovative IoT technology, and the programmers who are hacking information will discover it progressively hard to enter gadgets. The way toward going through the security level may turn out to be so tedious that it will be extremely simple for the specialists to get them.

8.7.1.2 Privacy

The process of information encryption is the vital component of blockchain. The utilization of bi-level cryptographic key guarantees that the process has an obvious spot to store information. The way that information cannot be gotten to with either the public or the private key and would expect to include together is the main benefit of having a blockchain. By combining this innovation being acquainted with IoT, smart gadgets will have the option to document the information of exchanges between them, which proves no conceivable method to release important data or control it. Besides, the information that goes through the blockchain is not able to alter. This makes it inconceivable for somebody to bargain the security capability of the IoT gadget being referred to.

8.7.1.3 No Centralized Data Warehouse Required

The apparent point of blockchain technology is the encryption guidelines that are unrivaled. It can also be confidently said that records are put away in a secured structure. In this manner, when the historical backdrops of all smart gadgets are encrypted in that same style, then there is no requirement for a centralized authorization to accumulate the entirety of the information. Herewith, an advanced level of trust is set up for overutilization in a traditional IoT technology.

The benefit of utilizing a blockchain is that the number of funds needed in the support of a distribution center should be possibly away totally. Therefore, the parties involved in this trade will remain to profit. The actual areas should likewise be possibly away with which can be a significant benefit that if it is considered for the bigger plan of things.

8.7.1.4 Smart Contracts

The essential idea of brilliant agreements is done by the encryption norms, which are actualized through blockchain and will be used to make contracts. The encryption rationale procedure will guarantee that the agreement is implemented during the predetermined rationales are fulfilled. Agreement terms and agreement conditions will be settled on commonly among both parties who are referred to. A safe and sound review is set up for this reason. Not exclusively will this idea of actualizing blockchain along with IoT be a distinct advantage in the realm of smart gadgets; however, it will likewise change how the business people direct business dealings. It should encourage enhanced data trade between the people who have associated with connected business measures also.

8.7.1.5   Industry Management

The convergence of this innovative technology has a boundless perspective in the business area. For example, any information in the oil industry will be greatly important. Thus, its abuse in the form of any malicious activity could represent a danger to public security. Using blockchain, information security can be guaranteed. Currently, the correspondence between smart gadgets could permit the network to consequently change its activities dependent on whether the data are received from the sensors to which it belongs. Subsequently, a traditional IoT organization can be enhanced by taking blockchain into the account. It helps to advance the general profitability of this oil business and could signify additional cash.

8.7.1.6   Supply Chain Management

The whole cycle of receiving food from Agri farmers to the racks of grocery stores is in the course of the supply chain, for instance. Along these lines, as the foodstuff, groceries are moving from the ranchers to providers and afterward to retailers through the mediators such as the market persons and wholesalers who are considered as hubs in the blockchain.

Presently, sensors used in IoT-based, the condition of the foodstuff, are observed at each progression. If any undesirable pesticides, bug sprays, or other shading specialist is included anyplace in the processing chain, it will promptly be recognized and the necessary steps can make a spot to guarantee that all the foodstuffs do not arrive at the client. Under the regular framework, it may get time of about 7 days to recognize the specific point in the supply chain in which the contamination happened. Using the coordinated effort of blockchain and the IoT, a similar issue can be identified with few seconds.

8.8   CONVERGENCE OF BLOCKCHAIN, AI, AND IOT

The convergence of blockchain, AI, and IoT [11] will produce a significant blend of features such as security, interconnectivity, and self-sufficiency to reform the state of affairs done (Figure 8.9).

The intermingling of blockchain, IoT, and AI can facilitate (Table 8.1) companies to increase the advantages of each technology when reducing the risk factors for each of its own. When IoT-connected devices are with various gadgets, positively, there are various possible malicious attacks by programmer assaults, extortion, and information robbery. To forestall security problems, AI coupled with machine learning

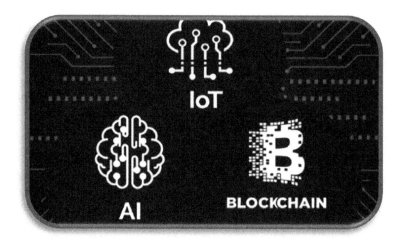

FIGURE 8.9   Convergence of IoT, AI, and blockchain.

TABLE 8.1
Sample Applications of Blending of Blockchain, AI, and IoT

Applications	Blockchain	Artificial Intelligence (AI)	Internet of Things (IoT)
Vehicles	Giving secure and permanent data source, which highlights confidential information for vehicles	Gather excellent features of vehicle	Improving fuel energy utilization for vehicles
Machinery	Giving secure and permanent data source, which highlights confidential information for machinery	Gather excellent features of machinery	Improving the procedure of manufacturing, production, and maintenance for machinery

technology will protect from hackers. Protecting data of a company can be additionally upgraded by blockchain, where restriction of unlawful access and changing company's information are completed limited.

AI can likewise upgrade the useful ability of the IoT system by enabling additional smart and self-sufficient features. Even though the normal effect of the intermingling of blockchain, IoT, and AI is energizing to consider, existing utilizations of these innovations are a long way commencing it ideal. Most of the organizations guarantee to have embraced AI, even though their functions are as yet in the beginning phases and not even close to the degrees of refinement needed to accomplish actual change. The equivalent analysis can be suitable for IoT and blockchain. Nonetheless, with the increased venture and development, the intermingling of blockchain, IoT, and AI will ultimately become real.

8.9 SECURITY ESSENTIALS FOR THE CONVERGENCE OF BLOCKCHAIN, AI, AND IOT

In the Internet, blockchain is reclassifying how solid exchanges are done through interconnected computers and settling the protection limitation experienced by IoT foundations and AI instruments. The majority of IoT gadgets are associated with one another using free and open connections, which might be hacked. Blockchain settles the security issue in these conditions, making direct records that are continually listed. The unified network technology like the client–server model is the one, much of the time utilized by IoT frameworks. This sort of foundation has far above the ground for supporting costs, as it utilizes common cloud frameworks and a huge network server that has host network connection tools. It is also considered as a contradiction in which there is a shared communication structure, which can give a powerful arrangement with regard to decreasing expenses, however with the related issue of an absence of safety. For this situation, blockchain cures this inadequacy by distributing and checking the exchanges through the hubs of the communication networks, as opposed to through a solitary common system of the server (Table 8.2).

- **Secret code development**: Cryptography [12] can be utilized to confirm and recognize the remaining hubs in that network and license them to append exchanges to the enormous blockchain crypt. Along with the innovative decentralization measures involved by the scattering of forces, capacities, and individuals, IoT gadgets will perform a significant function in progressing the group of information that is collected from this present reality to prepare the AI frameworks. In such a manner, protection and security will be the central point and the security organizations should make a solid effort to develop a blockchain framework that can satisfy these undertakings. The following security essentials are incorporated in the convergence of blockchain, AI, and IoT.
- **Digital identification number**: Digital identification number [13] is an arising strategy utilized for different IoT usages. Mostly, this gives a distinguishing

TABLE 8.2
Convergence and Performance of Trio Technology

Convergence of Blockchain, AI, and IoT	Performance of Trio Technology
Deep learning, machine learning, systematic intellectual	Smart devices utilized in health care, automobiles, agriculture, and home
Digital identification, function script, hash code	General-purpose and special-purpose procedure in business
Sharing cloud space, smart saving space, mini server	Potential on storage, decision-making, facility
Shared methods, decentralization	Error control and flow control, power utilization
Secret code development, validation, verification	Communication potential on a digital network
Information block, chain organization, bitcoin technology	Equipment, sensor, real-time devices, magnetic barcode, digital records

proof location to the gadgets, which communicate with one another progressively, safe, and decentralized mode. At this point, while some organizations, for example, Amazon and Flipkart, encourage to follow the ordered parcel from consignment to receiving, the digital identification number can be used by the bar code of each bundle.

In any case, the digital identification number is an essential idea for sensors, programmed control machines, and smart vehicles as well. Using the combination of innovative blockchain and AI for IoT, the digital identification number can be used to encoding the information in a computerized structure since it is a secret code; hence, anyone cannot utilize or hack from any place and at whatever point. This unique identification procedure can be utilized in administration and business to support the company's information in a carefully and novel structure. Utilizing by combining AI with blockchain for IoT, miniature Robo machines can be applied on insect killers to catch particularly for identifying the act of each plant. It gives monetary, operational focal points, for example, improved consistency in the harvest, lessens crops lost to sickness, the time, and exertion required, and diminishes the utilization and abuse of pesticides and herbicides.

By and large, digital identification is referring to people and organizations; however, it can likewise refer to IoT gadgets and machines. Blockchain-based digital identification will ensure that the exchange of gatherings will get a computerized digital code, which depends on their original physical identification such as for people identification cards and organizations their business register section. Using that type of digital identification, exchanges among people and organizations can be directed and handled productively with low exchange amounts and a high exchange speed.

- **Sharing cloud space**: The main centerpiece of blockchain innovation is sharing space [14], and it is utilized for putting away the data in a shared and decentralized structure in blockchain networks. These snippets of data can be checked by small people who adhere to the guidelines of a smart agreement. This strategy gives total map out, straightforwardness to the cloud in the style of AI technology. Shared storage gives another answer for an information storage issue that upgrades the size of information from IoT gadgets by the combination of blockchain and AI in IoT gadgets. Only for this cycle, the forecast idea in AI has a pivotal function in the expectation of shared information ahead of time and recommends using the information later on. Sharing cloud space idea utilized for offering explicit types of assistance, for example, database, decision-making to the IoT gadgets using shared cloud space in blockchain combined with IoT, smart agreement utilized in each square can perform exchanges in distributed designs without the utilization of a common central point by including an additional versatility and protection from digital assaults. This cycle likewise gives a safe, quicker, dependable method of correspondence for everyone in the network.

- **Decentralization**: IoT gadgets are interfacing using public organizations, which might be hacked by any third malignant client because IoT frameworks utilize concentrated worker systems. The utilization of blockchain innovation settles the security issue in IoT conditions and makes straight records that are listed constantly and are decentralized [15]. The assembly of blockchain and AI for IoT utilizes numerous agreement conventions, for example, verification of work, evidence of stake, assigned confirmation of stake and conveyance, decentralization instrument for versatility and security, the scattering of intensity utilized decentralization measure. IoT applications have a principal part in improving the assortment of information from different gadgets to prepare the AI framework. In this cycle, security, protection, and energy utilization are huge difficulties, with the union of blockchain, and AI tackles these issues.

- **Validation/verification**: Authentication is any technique; for example, a bank confirms a piece of the protected exchange starting with one individual and then onto the next individual; it is the idea utilized customarily in IoT applications, for example, clinical and smart vehicles. The IoT application utilizes a unified way in this cycle, and it is subject to the bank. Blockchain innovation gives the best approach to programmed exchange starting with one individual and then onto the next with cryptographically marked and checked by all excavators. It utilizes cryptographic money, for example, bitcoin, Ethereum for confirmation. AI is utilized for insightful and dynamic abilities for a particular exchange.

  The intermingling of blockchain and AI for IoT gives the engineering known as decentralized AI. It is utilized for a programmed exchange in a protected, true way and is confirmed by diggers. Confirmation and check are utilized in the correspondence between peer members. With the utilization of blockchain innovation in IoT, devices and entryways may ensure data, which is put away, prepared at the hub. All data are cryptographically checked in the circulated hyper record that is moved by all members hub, which can approve the trustworthiness before tolerating them.

- **Chain structure**: Chain structure [16] is an assortment of information in the IoT application. It was created by different detecting gadgets, for example, mobiles, Wi-fi, and pen drives. The actual layer or observation layer has a chain structure that is identified with the information base utilizing keen agreement, hash capacities, worldwide enrollment, and conveyed personality for blockchain framework capacity of IoT chain structure.

  The protection of information is particularly valuable with regard to IoT. In IoT, machines and gadgets store a high measure of delicate information. It is fundamental to guarantee protection and security of put away information. It is normal practice to send IoT information straightforwardly from the machine to the separate information base for assortment purposes. Be that as it may, this information does not have a serious level of protection as it is not encoded. Blockchain innovation adds an incentive as it can undoubtedly guarantee the security of the gathered information. In any case, there is a compromise between a significant level of security and control for unlawful exercises. If exchanges are mysterious, it is unimaginable to expect to gather the name and the address of the exchange sender. This namelessness highlights illegal exercises, for example, tax evasion or fear financing. For this situation, AI can be useful and can expand security by recognizing unlawful exercises. Note that AI advancements profit by the high measure of given IoT information as AI calculations gain from the information.

8.9.1 RECOMMENDATIONS

To adopt the benefits of all three innovative technologies and to accomplish prospective security features for any organization, there are few standard recommendations to follow by the strategy developers.

I. All innovation and technical development require financial support. Hence, government sector is required to look at such needs to finance for the researches on these newer technologies to obtain genuine advantages.

II. Considering this trio technology, additional consideration is required to advancing both best practice and capable responsibility for these advancements and the applications that are based on them. Governments could uphold these learnings and encourage research through such methods as advancement workplaces.

III. Specialists may need to reexamine and possibly update existing administrative and observing/reviewing measures to fulfill the needs of these new kinds of technologies.

IV. Private area commitment with strategy producers will likewise support more prominent interoperability just as information sharing, which is a vital aspect for accomplishing advancement at scale.

V. At every possible opportunity, governments should look to give administrative lucidity to trendsetters, what's more, put forth a solid attempt to plug and teach pioneers on the predominant administrative climate.

VI. While administrative lucidity is acceptable, as continually, managing excessively fast in arising advancements can prevent development. Strategy producers should hope to find some kind of harmony among insurance and advancement.

VII. There are huge moral contemplations in the utilization cases examined here, especially inside the setting of smart urban communities and other huge scope stages including individual information. These should likewise be featured for examination by strategy producers, who ought to hope to forestall undesired ramifications for society everywhere, and specifically powerless people and gatherings.

VIII. While making or adjusting administrative and lawful structures to cover exercises controlled by new and arising innovations, there should be an unmistakable concentration to manage the movement attempted, not the basic innovation. Innovation nonpartisan guideline can be more practical and is less inclined to hindering future advancement.

8.10 CONCLUSION

The convergence of blockchain, IoT, and AI is the development that gives enormous advantages to security by digital identification number and cryptography additionally straightforwardness, permanence, protection, and the automatization of company's measures. In any case, the effect of these trio developments is significantly higher, specifically with the help of sharing cloud space. These developments will combine to constrain the digitization of the business, in the future. This intermingling will expand the nature of information maintenance by arriving at a more significant level of normalization, protection, and securing the information by secret code development. In addition to that, novel plans of action are empowered with the end goal that self-sufficient specialists such as sensors, smart vehicles, smart machines, and smart cameras are to be decentralized. It can be placed as benefit points that self-govern transfer and get cash. It will be proposed to draw in with these innovative advances to acknowledge proficiency gains. Blockchain innovation converged with IoT and AI will prepare to provide another time of digitization.

REFERENCES

[1] Li, X., et al., (2020) "A survey on the security of blockchain systems", *Future Gen Comput Syst* 107, 841–853.
[2] Hou, Y., Garg, S., Hui, L., Jayakody, D. N. K., Jin, R., and Hossain, M. S. (2020) "A data security-enhanced access control mechanism in mobile edge computing", *IEEE Access* 8, 136119–136130.
[3] Vermesan, O., Broring, A., Tragos, E., Serrano, M., Bacciu, D., Chessa, S., Gallicchio, C., et al., (2017) "Internet of robotic things: converging sensing/actuating, hypoconnectivity, artificial intelligence, and IoT platforms", 97–155.
[4] Chu, S., and Wang, S. (2018) "The curses of blockchain decentralization", arXiv preprint arXiv:1810.02937.
[5] Benčić, F. M., and Žarko, I. P. (2018) "Distributed ledger technology: blockchain compared to directed acyclic graph", In *2018 IEEE 38th International Conference on Distributed Computing Systems (ICDCS)*, IEEE, pp. 1569–1570.
[6] Zhou, J., Wang, Y., Ota, K., and Dong, M. (2019) "AAIoT: accelerating artificial intelligence in IoT systems", *IEEE Wireless Commun Lett* 8(3), 825–828.
[7] Dai, H.-N., Zheng, Z., and Zhang, Y. (2019) "Blockchain for internet of things: a survey", *IEEE Internet Things J* 6(5), 8076–8094.
[8] https://www.statista.com/statistics/471264/iot-number-of-connected-devices-worldwide/.
[9] Wang, K., Dong, J., Wang, Y., and Yin, H. (2019) "Securing data with blockchain and AI", *IEEE Access* 7, 77981–77989.
[10] Merenda, M., Porcaro, C., and Iero, D. (2020) "Edge machine learning for AI-enabled IoT devices: a review", *Sensors* 20(9), 2533.
[11] Singh, S., Sharma, P. K., Yoon, B., Shojafar, M., Cho, G. H., and Ra, I. H. (2020) "Convergence of blockchain and artificial intelligence in IoT network for the sustainable smart city", *Sustainable Cities Soc* 63, 102364.
[12] Pasala, S., Pavani, V., Lakshmi, G. V., and Narayana, V. L. (2020) "Identification of attackers using blockchain transactions using cryptography methods", *J Crit Rev* 7(6), 368–375.
[13] AlQallaf, A. (2019) "Blockchain-based digital identity management scheme for field connected IoT devices", In *SPE Kuwait Oil & Gas Show and Conference*, Society of Petroleum Engineers.

[14] Sharma, P., Jindal, R., and Borah, M. D. (2020) "Blockchain technology for cloud storage: a systematic literature review", *ACM Comput Surv (CSUR)* 53(4), 1–32.

[15] Yousuf, S., and Svetinovic, D. (2019) "Blockchain trust and decentralization in supply chain management", In *2019 27th Telecommunications Forum (TELFOR)*, IEEE, pp. 1–4.

[16] Wu, J., Dong, M., Ota, K., Li, J., and Yang, W. (2020) "Application-aware consensus management for software-defined intelligent blockchain in IoT", *IEEE Network* 34(1), 69–75.

9 Convergence of Blockchain and Artificial Intelligence in IoT for the Smart City

A. Peter Soosai Anandaraj
Veltech Rangarajan Dr. Sagunthala R&D
Institute of Science & Technology

A. Ilavendhan
Veltech Rangarajan Dr. Sagunthala R&D
Institute of Science & Technology

P. Kavitha Rani
Sri Krishna College of Engineering and Technology

R. Sendhil
Madanapalle Institute of Technology & Science

CONTENTS

DOI: 10.1201/9781003081180-9

9.1   INTRODUCTION

The rapid urbanization of the world's population produces a slew of economic, environmental, and social issues that have a substantial impact on many people's lifestyles and quality of life. With the high population density of metropolitan zones, the implementation of the smart city offers chances to overcome these issues and deliver a better lifestyle through high-quality intelligent services. The eco-friendly smart cities concept aims to create knowledge and advertise green energy best practices with little usage. According to the United Nations, 66% of the world's population [1] will soon live in cities, implying that we will confront substantial challenges in terms of social sustainability. Furthermore, the modern city's form is considered a social and environmental issue. Municipalities absorb nearly 70% of the world's resources, posing considerable obstacles to the distribution of these resources using cutting-edge technologies [2]. In the deployment of smart cities, information and communication technology (ICT) plays a critical role. Many complementary aspects of emerging technologies, such as blockchain technology and Artificial Intelligence (AI), include trustlessness, automation, decentralization, democratization, and security. Since the term was coined in 2008, interest in blockchain technology has grown; the concept is gaining traction and can be considered a hot issue [3]. Blockchain is a distributed network's decentralized, public ledger of all transactions. Protecting against cybercrime, traditional crime, natural catastrophes, or terrorism, public safety, and security has become a top priority for smart city governments. According to IIoT world, nearly 700 million dollars was invested on Internet of Things (IoT) privacy and security issues in 2017, with that number expected to climb to 4.4 billion dollars by 2022 [4].

When it comes to smart city security and privacy issues, it's important to note that many of the complex issues persisting now will not happen, when blockchain, AI, and IoT technologies are integrated. To address the security and privacy difficulties that smart cities confront, stakeholders must come together as a group to ensure that certain problems do not spread throughout the intelligent network. To attain this lofty goal, motivated security professionals and smart city planners must take advantage of existing initiatives and sustainable cities to support the development of the necessary smart technologies. The current thoughts and attempts to address these security risks and challenges will influence how future smart sustainable towns are created.

Furthermore, blockchain is one of the safest and most dependable architectures for constructing a newly built parallel intelligence transport management system. Blockchain is evolving at a breakneck pace, and it has the potential to revolutionize the way Intelligent Transportation Systems (ITSs) are used to create sustainable smart cities. We can make better use of current ITS infrastructure and resources, especially for crowdsourcing technology, by integrating blockchain to construct a safe, dependable, and decentralized autonomous transportation system ecosystem. In the not-too-distant future, ITS will be made up of a huge network of self-driving vehicles in smart cities.

Even in our closed society, demand for ITSs is growing since the ease of private mobility comes at a significant cost in terms of road resource use. Human aspects are important in ITS, but they are also important in other domains of technology where ITS will have a significant impact, such as vehicle control design and traveler information management [5,6]. The transportation industry is the legal means of moving commodities from one location to another. Transportation faces numerous issues over time, including high accident rates, carbon emissions, air pollution, and traffic congestion. Furthermore, the open nature of ITS, as well as its usage of wireless communication technology, raises numerous security and privacy concerns. Integrity, confidentiality, location privacy authentication, nonrepudiation, identity privacy, anonymity, certificate revocation, and certificate verification are among the issues that must be addressed.

Although blockchain technology has major implications for future smart network environments, it also confronts several technical obstacles, many of which are connected to security and privacy, trust, scalability, and business weaknesses. Several researchers have focused on smart contract vulnerabilities, such as those found in Ethereum, bitcoin, and wallet contracts [7], as well as the security difficulties that go along with them [8]. The IoT has emerged as a potentially large impact area with the introduction of smart networks such as smart homes and cities; 50 billion linked devices are projected by 2020. Over the last few years, blockchain has played a key role in enhancing IoT connections. A distributed blockchain architecture for future smart city networks is depicted in Figure 9.1. The distributed ledger system provides for monitoring by keeping track of which devices (such as smart automobiles) are connected and what action is taken in each encounter. Smart contracts can be used by IoT devices to communicate with one another and connect to the Internet, eliminating the requirement for a central authority to validate any contact between two parties [9]. Large organizations may employ IoT and blockchain to improve their security architecture, as well as to gather and analyze data.

Open ledgers can facilitate fault-proof integration of connected devices in smart networks with blockchain IoT solutions, removing obstacles like data transparency and end-to-end process tracking.

- Automating and verifying transactions
- Real-time data exchange throughout the network
- The stress-free scalability of cloud-based IoT platforms is a restriction.
- Data analytics for IoT and network challenges

9.2 SMART CITIES

Cities are quickly expanding over the world; it is predicted that two-thirds of the world's population will be living in cities by 2050. This signifies that cities are expanding in terms of population and size. Such expansion magnifies the scale of

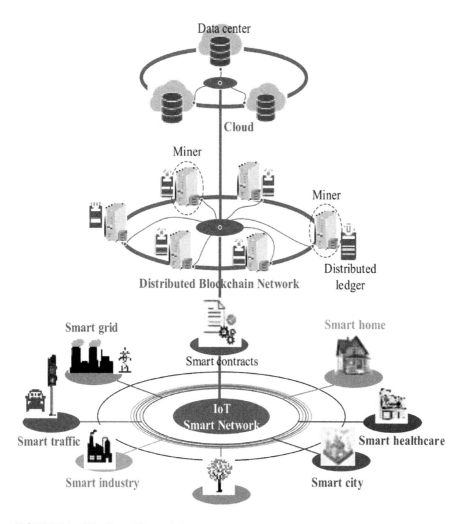

FIGURE 9.1   Distributed blockchain-based smart city network architecture.

current concerns in cities or introduces new challenges. For example, pollution, congestion, waste management, and energy needs will continue to rise [10]. Legacy ways to solving such challenges will not produce efficient results, even if they can overcome the basic issues [11].

One of the key drivers driving current thinking and the development of smart city concepts appears to be tackling city expansion challenges. Cities will create 80% of economic creation and 60% of energy use, according to MIT research published in 2018. More innovative people will be drawn to the city as the population grows. As a result, smart cities could become a platform for entrepreneurs.

Focusing on the components and requirements for converting smart city platforms from providing digital services to platforms that enable community members to innovate and develop applications that are meaningful and have greater value is an intriguing perspective on the benefits that smart cities can provide [12,13].

Smart cities, in general, can provide several tangible benefits [14,15]:

- The protection of citizens. Intelligent surveillance is one example of a technology that can be used.
- High-tech medical care.
- ITSs and smart mobility are two topics that come up frequently. Such practical examples include ride-sharing and smart parking.
- Controlling energy consumption. Examples are renewable energy and real-time energy monitoring.

9.2.1 CHALLENGES TO BUILDING SMART CITIES

This section will cover the strategic, platform, and technological problems that come with creating smart city environments. The next sections will outline current solutions and technological platforms that may be used to address those issues.

One of the most underappreciated yet critical issues in establishing a smart city is defining what a smart city means to residents and what problems it will address for them [] stated, for example, that the inhabitants of Barcelona were not involved in developing or understanding the notion of smart city. This suggests that Barcelona people are unaware of the advantages of living in a smart city.

The transition of smart city ideas into actual programs that drive change and produce particular results is another nontechnical problem [16]. When introducing the smart city program, every city must have a vision, strategic themes, and goals. Defining a smart city implementation strategy necessitates collaboration across a vast number of stakeholders, departments, businesses, and services, making agreement even more difficult. To analyze the validity and sustainability of smart city projects, they must be linked to strategic objectives. To establish this connectivity, the city must take a deliberate approach to bridge the gap between the strategy and the objectives represented in projects.

9.2.1.1 Security

The risk of a data breach is quite significant when it comes to the acquisition of personal data that is necessary for offering IoT and WSN-based solutions for smart cities. When personal and privileged information is made public, the possibilities of it being misused or stolen increase dramatically [17]. There have been countless occasions where cyber hackers have attempted to access information held on servers. This stolen information can be used to access bank accounts, personal and private information, social security numbers, and anything else in the database [18,19].

Confidentiality, integrity, and availability are all important aspects of security. This is no different from the security of smart cities, where lax security leads to criminal actions, data integrity difficulties, invasion of privacy, service disruptions, and a slew of other issues. All stakeholders, including citizens, government institutions, and corporations, are directly affected by such hazards. AlDairi [20] divides security concerns into two categories: infrastructure security and data privacy, which are comparable to the classification suggested by Khatoun in 2016. Infrastructure is vulnerable to a wide range of security risks.

Because the smart city relies on connectivity between devices, such as wireless connections, such connections are vulnerable to Denial of Service (DoS) assaults, which can result in availability difficulties. Cerrudo (2015) provided a comprehensive overview of infrastructure-related security issues [21], of which we will highlight a few: (i) insufficient security testing, (ii) inadequate security in software and IoT devices produced by many providers, (iii) insecure encryption management – particularly in the generation and maintenance of encryption keys, which is a major issue in a system that relies on wireless connectivity, such as smart cities, (iv) inadequate patch management methods, which are mostly due to the challenges of testing patches in nonproduction environments and vendor delays in issuing updates, (v) Relying on old, out-of-date systems, such as running the backend of a building management system on an out-of-date operating system. Simple vulnerabilities could have a significant impact on citizens' daily lives in a smart city if security is not properly managed; for example, a vulnerability in a smart parking management system that displays available parking locations in the city could cause havoc by causing the system to show that most on-street parking spaces are available while they are full during peak hours.

Data privacy is another facet of security concerns. One of the most important aspects of this issue is data ownership: if data is owned by private companies, security and privacy are less of a concern (for example, they may sell the data to third parties); however, if the data is owned by government entities, personal information privacy and security are more of a concern. Many smart cities, for example, would rely on cloud solutions to collect and store data from various sources such as sensors, IoT devices, cameras, and so on.

9.2.1.2   Accountability

Governing bodies that compile and store data generated by IoTs and WSNs often fail to provide accountability to citizens by failing to demonstrate how the data is being used to improve residents' quality of life. The data can be perplexing at times, making it difficult to determine how to use a specific collection of data, which is the basis of the problem.

Data management: In a smart city, there is a great amount of data from many sources and types that require real-time collection, processing, storage, aggregation, analysis, access, and security. Sensors, mobile apps, and various forms of software are among the data sources, with sensors serving as the primary data source. Sensors globally produced 1250 billion GB of data in 2010, according to Baraniuk (2011) [22].

Sensor data, on the other hand, has the following characteristics: high velocity, high variability, and high volume necessitating the development of platforms and technology that can manage these issues. Aside from those data requirements, the biggest source of concern is probably managing and analyzing the collected data in a way that is easy to communicate and deliver value to decision-makers and other customers.

9.2.1.3   Interoperability

At different tiers of the smart city architecture, this difficulty can take on diverse forms. This issue could, for example, be related to the lack of standard or open protocols to facilitate interoperability and communication among various types and manufacturers of IoT devices at the infrastructure layer [23]. On the other hand, there are no standards for interacting with IoT devices at the upper levels where applications are supposed to interface with IoT devices [24]. Application developers, for example, will be required to construct interfaces that link with IoT product-specific Application Programming Interfaces (APIs), resulting in silos and complex structures.

9.2.1.4   Open Data

The notion of Open Data, which refers to free and open access to government data, is gaining traction in many places. A smart city is a repository for a significant amount of data gathered from many sources. When it comes to providing open access to data, cities face many issues, including security, privacy, data formats, data management, and data access.

9.2.1.5   Budget

It cannot be argued that implementing new technology is a costly endeavor that necessitates significant investments in procurement and infrastructure development. Residents, who may be hesitant to support this cause, must ultimately bear the burden of the impending cost [25]. Another issue that is being intensively researched is security and privacy.

9.2.1.6   Solutions

Smart cities will follow in the footsteps of industries like healthcare, education, retail, and agriculture, where technology has played a key role. According to Pablo and colleagues [26], technology is the backbone of an intelligent city. As a result, various technologies that might help manage and operate a smart city are discussed below. These sophisticated and new technologies' integration and application in smart cities were also explored, as well as the difficulties and solutions they confront.

9.3   CONVERGENCE OF BLOCKCHAIN, ARTIFICIAL INTELLIGENCE, AND IOT

With the advancement in the field of technology, we have seen the emergence of various new terms in the market. The advancement made in the field of AI, the IoT, and blockchain has coined the term, "*smart city*". A smart city is a municipality-based framework that is predominantly controlled or composed of ICT. The main idea of the introduction of the term "smart city" was to increase operational efficiency and to help in upgrading the quality of life in the urban city.

The main mission that is aimed behind the smart city is to optimize the city functions and provide economic growth as well as improving the quality of life for its citizen using smart technologies as well as data analytics. It is a misconception in people that a smart city is valued on how many technologies a city uses. Though it is not correct, the value of the smart city is based on the fact that how the technologies are used to provide ease in the living of the citizens.

The ICT, which is the backbone of the smart city, usually comprises intelligent networks. These networks are connected objects and machines that communicate with each other by transmitting data with the help of wireless technology and cloud services. Citizens residing in the smart city communicate with the ecosystem by using smartphones, mobile devices, and connected cars and homes. Communities in a smart city can help in energy distribution, streamlining trash collection, decreasing traffic congestion, and also improving air quality, and all are done with the help of the IoT.

All these are combined, thus making the term "smart city". These technologies amalgamate into the term known as "smart city" (Table 9.1).

9.4   ARTIFICIAL INTELLIGENCE AND BLOCKCHAIN

As shown in Table 9.2, AI [27–31] and blockchain [32–34] technologies contradict each other when used alone and operate in completely distinct paradigms, but combining the two is promising for a variety of industries.

AI and blockchain integration is the convergence of two of the most cutting-edge technologies, which have the potential to offer up a wide range of options. ITSs for self-driving automobiles may be the finest opportunity. The concept is for a driverless automobile system, self-learning cars, and a smart city transportation system to use an artificially intelligent blockchain. Autonomous vehicles have become commonplace, but not everyone can afford one. One option is to rent a car. By lowering the number of individuals involved, syncing data, and providing audit trails, blockchain technology can help to streamline the renting process.

Any intelligent car system will gain greatly from this. Heterogeneous vehicles interact and share experiences effectively using blockchain and AI.

TABLE 9.1
Working Definitions of Smart Cities

References	Definition
Giffinger & Gudrun (2010, 2018)	"A city that excels in a variety of ways in the future, founded on a clever mix of endowments and activities of self-determined, independent, and aware citizens".
Hall (2000)	"A city that keeps track of and combines the state of all of its vital infrastructure".
Hartley (2005)	A city with "physical infrastructure, IT infrastructure, social infrastructure, and commercial infrastructure connected to exploit the city's collective intelligence".
Toppeta (2010)	A city that "combines ICT and Web 2.0 technology with other organizational, design, and planning efforts to de-materialize and speed up bureaucratic processes while also assisting in the identification of new, innovative solutions to city management complexity to improve sustainability and liveability".
Washburn et al. (2010)	"The application of Smart Computing technology to make a city's vital infrastructure components and services more intelligent, networked, and efficient". This includes city administration, education, healthcare, public safety, real estate, transportation, and utilities.

TABLE 9.2
Various Aspects between Blockchain and AI

Features	Blockchain	Artificial Intelligence
Character	Decentralized	Centralized
Access	Open	Closed
Transparency	Transparent	Black box
Perspective	Deterministic	Probabilistic

- Time savings, cost savings, and a reduction in human effort
- Ensure that the learning process is modular.
- Increasing efficiency by minimizing the number of cars that must be trained separately

The major drivers of today's innovation are AI and blockchain. Both are projected to completely transform our lives and contribute trillions of dollars to the global economy. On the one hand, the blockchain has problems in terms of scalability, security, and efficiency; on the other hand, AI has its own set of privacy and trustworthiness concerns. The amalgamation of these two technologies will work together to alter the world. The blockchain can give security and privacy. Machine learning algorithms can be built on the blockchain to gain security, scalability, and trustworthiness.

As a result, AI and blockchain integration may be thought of as the blockchain for AI and the AI for blockchain. Some of the key issues to be addressed in the convergence of blockchain and AI include security and privacy, threats and attacks, intelligent infrastructure, technical and business challenges, a lack of standards, interoperable regulations, smart contract vulnerabilities, and deterministic executions, as well as good governance.

9.5 SOLUTIONS FOR BLOCKCHAIN-AI-BASED SMART CITIES

For car ad hoc networks, Benjamine et al. [35] discussed blockchain technology (VANET). For a car ad hoc network, they integrated the Ethereum blockchain with a smart contract framework. They recommended combining two sorts of applications: mandatory and optional. To deliver the VANET service, they attempted to connect utilizing the blockchain. The blockchain can be utilized for a variety of purposes, including vehicle communication, security, and peer-to-peer communication without revealing personal information. Ali Dorri [36] developed a blockchain technology mechanism for providing and updating wireless remote software without disclosing personal information about vehicles. In a peer-to-peer network, Madhusudan Singh and Shiho Kim [37] established a framework for intelligent vehicle data exchange. They pointed out that vehicle communication application security standards are outmoded, relying on common mobile phone and computer security procedures that are more suitable for ITS applications. Many academics are still attempting to develop standard transportation security and efficiency procedures [38–41].

Sean Rowan et al. [42] explored the visibility of autonomous and self-driving automobiles on public highways. They proposed employing visible sidelights (with a CMOS camera) and acoustics (ultrasonic sounds) to code. The side channels are studied both theoretically and practically, and an upper limit for the rate of line code modulation that may be reached with these side-channel techniques in a vehicle network is established. A novel vehicle session key establishment mechanism was proposed, which utilizes both side channels and a blockchain public key architecture.

Ghulam Mujtaba and Nadeem Javaid highlighted how ITS will have a large network of autonomous cars shortly, including ambulances, law enforcement vehicles, public transportation, domestic automobiles, and so on. These vehicles will become a part of the network and will contribute entirely to its efficiency. However, in the case of fleets of any kind that operate in groups and work for an organization, these vehicles must be studied to ensure that they are compatible with ITS. They developed a system in which fleet vehicles are connected to a single ITS network that offers services to all autonomous vehicles while they go about their business [43].

The combination of blockchain and AI might usher in the next industrial revolution. An AI might be used to deliver bug-free smart contracts, opening up a slew of new opportunities. The following are some of the ways AI can help with blockchain deployment.

Sustainability: AI can assist the blockchain system in reducing energy consumption.

Scalability: Without centralized datasets, AI algorithms may learn from remote data sources and engage in collaborative learning.

Security: IDS/IPS issues in blockchain can be detected by AI. More resilient ciphers are provided by computational intelligence, which improves blockchain resilience systems.

Privacy: Intelligent search techniques can increase the performance of hash functions.

Hardware: To improve system performance, AI can be used to improve the design of mining hardware.

AI is currently infiltrating our lives in a variety of unexpected ways. The usage of smart contracts in conjunction with the blockchain helps to limit any AI-related issues. For example, a smart contract-encoded law might put a stop to rule-breaking in self-driving automobiles.

9.6   IMPACT OF THE INTERNET OF THINGS IN A SMART CITY

The following are some of the most notable and crucial functions that IoT performs in a smart city.

9.6.1   TRAFFIC MANAGEMENT

In a smart city, IoT implementation has enormous promise for regulating traffic signals and road traffic. Data obtained from a grid of cameras, other equipment, and sensors can be used to analyze traffic flow, congestion, and jam-prone streets, as well as the impacted streets and roads during rush hours and the duration of traffic lights on a busy street [44]. The information gathered can be utilized to manage, adjust, and guide traffic signals such that congestion and waiting times are reduced significantly. In all smart cities around the world, the outcomes of implementing the technology have been spectacular. The Los Angeles, city in the United States, has invested $400 million in a system that uses photographic cameras and sensors embedded on the road to deliver the real-time data to a central computer system to control and manage traffic. The system is in charge of 4500 traffic signals throughout the city. Since its completion in 2013, the new system has been credited with 16% faster traffic and a 12% reduction in waiting time and congestion.

9.6.2 WATER SUPPLY MANAGEMENT

In most cities, water supply and management are a long-standing tradition. Water is stored, pumped, and delivered to clients using outdated infrastructure and resources. This aging system is primarily reliant on manual labor, making fault and defect detection extremely challenging. To report a problem, utility firms rely on client complaints. As a result, there is a great deal of waste, expensive operating and maintenance costs, and a delay in responding to a situation. Using IoT to monitor pipes and detect and report issues, as well as to register usage and customer complaints, makes water management far more productive and cost-effective. Without manual input, the defect detection process can become nearly error-free. The potential for IoT in water management is enormous, especially considering the resource's increasing scarcity and governments throughout the globe devising conservation programs as new and less expensive technology becomes widely available. With the understanding that water is the most essential resource for humans and will only grow scarce as a result of imminent climate change, water management is a major concern for IoT in smart cities across the world. Approximately 20% of water expenses can be reduced effectively using IoT in a smart city [45].

9.6.3 TRANSPORTATION

Integrating, controlling, and expanding a smart city's transportation are critical since it has a wide range of implications for overall growth and resource accessibility. To offer a strong and eco-friendly environment for its inhabitants, a smart city must prioritize contemporary and seamless public transit. This would not only enhance accessibility across the city but will also reduce pollution and traffic congestion triggered by the millions of automobiles that utilize the roads daily. The way forward is to provide citizens with affordable, dependable, and conveniently accessible alternate modes of transportation. The IoT offers assorted tools to derive solutions to assist a smart city to rectify its issues related to transportation. One such product that makes travel easier is smart Radio-frequency identification (RFID)-implemented cards that unite several means of transportation. Commuters use RFID technology to board a bus, tram, or metro train, instead of purchasing separate tickets for each service. Using apps and online portals to purchase tickets or recharge smart cards reduces the need to queue at ticket vending machines or counters. When all of these factors are considered, we may expect a significantly pleasant travel experience that saves money, time, and emissions. As a result, more commuters will use public transportation, and the ever-evolving nature of technology will achieve the intended effects for all citizens in terms of a healthy and sustainable environment [46]. For example, Chicago in the United States has introduced integrated Ventra Card payment for its transit services. Commuters can use the card to travel on Chicago Transit Authority (CTA), Pace, and Metra services, eliminating the need to purchase multiple tickets.

9.7 CONCLUSIONS

The smart city industry is transforming as a result of the blockchain, AI, and IoT convergence paradigm that brings businesses, governments, and even countries together. Due to its decentralized structure and peer-to-peer qualities, blockchain, AI, and IoT technology is well known and highly appreciated. We explored the many problems in several parts of sustainable smart cities in this study, which highlighted the various security issues and challenges that limit the adoption of blockchain-AI technology. In addition, the poll looked at several active blockchain-AI initiatives and their commercial advantages. Finally, we explored blockchain security solutions for smart city transportation systems, as well as concerns of sustainability and presence in life cycle studies based on blockchain-AI and IoT.

REFERENCES

[1] United Nations, Department of Economic and Social Affairs, New York. (2015) http://esa.un.org/unpd/wup/Publications/Files/WUP2014-Report.pdf.

[2] Bibri, S. E., and Krogstie, J. (2017) "Smart sustainable cities of the future: an extensive interdisciplinary literature review", *Sustainable Cities Soc* 31, 183–212.

[3] Nakamoto. (2008) "Bitcoin: A Peer-to-Peer Electronic Cash System", https://bitcoin.org/bitcoin.pdf.

[4] https://iiot-world.com/reports/an-overview-of-the-iot-security-market-report-2017-2022.

[5] Sadek, A. W. (2007) "Artificial intelligence applications in transportation", In *Transportation Research Circular E-C113: Artificial Intelligence in Transportation: Information for Application*, Transportation Research Board of the National Academies, Washington, D.C.,, pp. 1–7.

[6] Li, X., Jiang, P., Chen, T., Luo, X., and Wen, Q. (2016) "A survey on the security of blockchain systems", *Future Gen Comput Syst* 9604, 106–125.

[7] Zhang, L., Zhipeng, C., and Xiaoming, W. (2016) "FakeMask: a novel privacy preserving approach for smartphones", *IEEE Trans Network Serv Manage* 13(2), 335–348.

[8] Size of the Blockchain Technology Market Worldwide from 2016 to 2021. (2018) https://www.statista.com/statistics/647231/worldwide-Blockchain-technology-market-size/.

[9] Singh, S., Sharma, P. K., Yoon, B., Shojafar, M., Cho, G. H., and Ra, I.-H. (2020) "Convergence of blockchain and artificial intelligence in IoT network for the sustainable smart city", *Sustainable Cities Soc* 63, 102364. doi: 10.1016/j.scs.2020.102364.

[10] Hanes, D., Salgueiro, G., Grossetete, P., Barton, R., and Henry, J. (2017) *IoT Fundamentals: Networking Technologies, Protocols, and Use Cases for the Internet of Things*, Indianapolis, IN: Cisco Press.

[11] GascÓ-Hernandez, M. (2018) "Building a smart city: lessons from Barcelona", *Commun ACM* 61(4), 50–57.

[12] Apolinarski, W., Iqbal, U., and Parreira, J. (2014) "The GAMBAS middleware and SDK for smart city applications", In *IEEE International Conference on Pervasive Computing and Communication Workshops (PERCOM WORKSHOPS)*.

[13] Wu, C., Birch, D., Silva, D., Lee, C., Tsinalis, O., and Guo, Y. (2014) "Concinnity: a generic platform for big sensor data applications", *IEEE Cloud Comput* 1(2), 42–50.

[14] Khatoun, R., and Zeadally, S. (2016) "Smart cities", *Commun ACM* 59(8), 46–57.

[15] Cosner, S., and Zeadally, S. (2018) "Smart city solutions", *Public Manage* 00333611.

[16] GascÓ-Hernandez, M. (2018) "Building a smart city: lessons from Barcelona", *Commun ACM* 61(4).

[17] Bastidas, V., Bezbradica, M., and Helfert, M. (2017) "Cities as enterprises: a comparison of smart city frameworks based on enterprise architecture requirements", *Smart Cities* 20–28.

[18] Lacinák, M., and Ristvej, J. (2017) "Smart city, safety and security", *Procedia Eng* 192, 522–527.

[19] Braun, T., et al., (2018) "Security and privacy challenges in smart cities", *Sustainable Cities Soc* 39, 499–507.

[20] Eckhoff, D., and Wagner, I. (2018) "Privacy in the smart city—applications, technologies, challenges, and solutions", *IEEE Commun Surv Tutorials* 20(1), 489–516.

[21] Khatoun, R., and Zeadally, S. (2016) "Smart cities", *Commun ACM* 59(8), 46–57.

[22] AlDairi, A. (2017) "Cyber security attacks on smart cities and associated mobile technologies", *Procedia Comput Sci* 109, 1086–1091.

[23] Cerrudo, C. (2015) "An emerging US (and world) threat: cities wide open to cyber attacks", *Securing Smart Cities* 17, 137–151.

[24] Baraniuk, R. (2011) "More is less: signal processing and the data deluge", *Science* 331(6018), 717–719.

[25] Santana, E., Chaves, A., Gerosa, M., Kon, F., and Milojicic, D. (2017) "Software platforms for smart cities", *ACM Comput Surv* 50(6), 1–37.

[26] Soursos, S., Zarko, I., Zwickl, P., Gojmerac, I., Bianchi, G., and Carrozzo, G. (2016) "Towards the cross-domain interoperability of IoT platforms", In *European Conference on Networks and Communications (EuCNC)*.

[27] Silva, B. N., et al., (2018) "Towards sustainable smart cities: a review of trends, architectures, components, and open challenges in smart cities", *Sustainable Cities Soc* 38, 697–713.

[28] Pablo, C., et al., (2018) "Tendencies of technologies and platforms in smart cities: a state-of-the-art review", *Wireless Commun Mobile Comput* 2018.

[29] B-IoT: Blockchain Technology for IoT in Intelligent Transportation Systems, http://iot.ed.ac.uk/projects/b-iot/.

[30] Wang, D., Huang, L., and Tang, L. (2017) "Synchronization criteria for discontinuous neural networks with mixed delays via functional differential inclusions", *IEEE Trans Neural Networks Learning Syst* 29(5), 1809–1821.

[31] Zeng, D., Dai, Y., Li, F., Sherratt, R. S., and Wang, J. (2018) "Adversarial learning for distant supervised relation extraction", *Comput Mater Continua* 55(1), 121–136.

[32] Wang, X. H., He, Y. G., and Li, T. Z. (2009) "Neural network algorithm for designing FIR filters utilizing frequency-response masking technique", *J Comput Sci Technol* 24(3), 463–471.

[33] Sgantzos, K., and Grigg, I. (2019) "Artificial intelligence implementations on the blockchain: use cases and future applications", *Future Internet* 11(8), 170–182.

[34] Kefa, R. (2018) "Convergence of AI, IoT, big data and blockchain: a review", *Lake Inst J* 1(1), 1–18.

[35] Makhija, P. (2019) Top 5 Reasons: How Blockchain and Artificial Intelligence is Revolutionizing the Transport Industry, https://it.toolbox.com/guest-article/top-5-reasons-how-blockchain-and-artificial-intelligence-is-revolutionizing-the-transport-industry.

[36] Marwala, T., and Xing, B. (2018) "Blockchain and artificial intelligence", arXiv preprint arXiv:1802.04451, 1–12.

[37] Leiding, B., Memarmoshrefi, P., and Hogrefe, D. (2016) "Self-managed and blockchain-based vehicular ad-hoc networks", In *Proceedings of the 2016 ACM International Joint Conference on Pervasive and Ubiquitous Computing: Adjunct (UbiComp '16)*, ACM, New York, NY, pp. 137–140.

[38] Dorri, A., Steger, M., Kanhere, S. S., and Jurdak, R. (2017) "Blockchain: A distributed solution to automotive security and privacy", eprint arXiv:1704.00073.

[39] Singh, M., and Kim, S. (2017) "Blockchain based intelligent vehicle data sharing framework", arXiv preprint arXiv:1708.09721.

[40] Zhou, S., Liang, W., Li, J. and Kim, J. U. (2018) "Improved VGG model for road traffic sign recognition", *CMC Comput Mater Continua* 57(1), 11–24.

[41] Xiang, L., Li, Y., Hao, W., Yang, P., and Shen, X. (2018) "Reversible natural language watermarking using synonym substitution and arithmetic coding", *Comput Mater Continua* l(3), 541–559.

[42] Huang, Y. S., and Wang, Z. Y. (2014) "Decentralized adaptive fuzzy control for a class of large-scale MIMO nonlinear systems with strong interconnection and its application to automated highway systems", *Inf Sci* 274, 210–224.

[43] Qin, Z., Li, H., and Liu, Z. I. (2014) "Multi-objective comprehensive evaluation approach to a river health system based on fuzzy entropy", *Math Struct Comput Sci* 24(5), 1–8.

[44] Rowan, S., Clear, M., Huggard, M., and Goldrick, C. M. (2017) "Securing vehicle to vehicle data sharing using blockchain through visible light and acoustic side-channels", eprint arXiv:1704.02553.

[45] Mujtaba, G., and Javaid, N. (2020) "Blockchain based fleet management system for autonomous vehicles in an intelligent transport system".

[46] Masek, P., et al., (2016) "A harmonized perspective on transportation management in Smart cities: the novel IoT-driven environment for road traffic modeling", *Sensors* 16(11).

[47] Bellias, M. (2017) IoT for Water Utilities, from https://www.ibm.com/blogs/internet-of-things/iot-for-water-utilities.

[48] Masek, P., et al., (2016). "A harmonized perspective on transportation management in smart cities: the novel IoT-driven environment for road traffic modeling", *Sensors* 16(11).

[49] Caywood, M. (2018) How IoT Enabled Infrastructure Will Revolutionize Public Transit, from https://www.iotevolutionworld.com/smart-transport/articles/438023-how-iot-enabled-infrastructure-will-revolutionize-public-transit.htm.

10 Blockchain for Supply Chain in Pharmaceutical Industry

An Approach to Counterfeit Drug Detection

Kavita Saini and Kavita Kumari
Galgotias University

Deepak Kumar Saini
ST Microelectronics

CONTENTS

DOI: 10.1201/9781003081180-10

10.1 INTRODUCTION OF BLOCKCHAIN

Blockchain is termed as a highly reformed technology that is supportive to achieve a lot in the future. Certain features of blockchain enable secured recording and sharing of delicate information over the Internet. One such feature is distributed ledger technology (DLT). Several blockchain platforms are developed and adapted over time in the real world. The classification of these applications can be done in the following three sorts public, private, and consortium blockchain [4,5]. This chapter gives the details of a basic introduction to blockchain and its functionalities in healthcare and an overview of counterfeit drugs [7,9]. Figure 10.1 depicts the blocks used to form a Blockchain.

10.2 INTRODUCTION TO PHARMA SUPPLY CHAIN

The pharmaceutical business is a mind-boggling undertaking loaded with clashing goals and various unmanageable imperatives [1]. A profoundly directed condition combined with the life-changing nature of the items portrays the pharmaceutical business as a particularly troublesome framework [4,7]. A basic survey will recommend that pharma supply chain-related issues are not liable to figure among the greatest difficulties confronting the pharmaceutical business (see Figure 10.2). For a multi-billion sector that fabricates and appropriates items to a large number of individuals consistently, neglecting to see supply chain issues surely appears to be strange and worth examining. Figure 10.2 depicts the frequently cited pharmaceutical industry issues in the form of bar chart.

Interestingly, organizations, for example, Walmart, Dell, and Amazon.com have made it to the top absolutely on the quality of the supply chains they have. Truth be told, visionary supply chain board arrangements and advancements in information technology are clearing the market. A proficient and coordinated supply chain is currently viewed as basic for building up a manageable upper hand. All in all, what separates the pharmaceutical business from different divisions in such a manner?

Without a doubt, there are some conspicuous clarifications for the unreasonable conduct. All things considered, the pharmaceutical business' beneficial legacy is liable for its absence of spotlight on supply chain efficiencies [4,7]. Another chance is that the general ease of goods sold makes it simple for the administration to pick a dim-witted procedure of buffering all issues with stock. As it were, apparently there are practically zero catalysts to appropriately address inside organization efficiencies

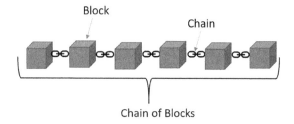

FIGURE 10.1   Blockchain blocks.

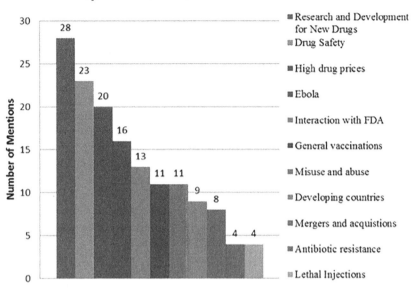

FIGURE 10.2 Most frequently cited pharmaceutical industry issues.

in the pharmaceutical business. So, for what reason would it be a good idea for us to explore the pharmaceutical business supply chain?

10.3 PHARMA SUPPLY CHAIN CONDITION IN INDIA

Chennai is today known as a center point for bleeding-edge clinical consideration, with emergency clinics, for example, Aravind Eye Hospitals and Apollo Hospitals, and nearly 42,000 drug stores making it an operational hub for both frontline and moderate treatment for patients around the globe. Barely 10 years prior, in 2010, the city's clinical and pharma accreditations endured a terrible shot, when a fake medication racket was uncovered and its boss Meenakshi Sundaram, a customer with the city's wash set, was outed as the pioneer of these phony drug rings. These fake medications had deadly outcomes – a child was slaughtered when given fake prescriptions, and the administration was shocked to take immediate actions, such as reporting clearing attacks and capturing numerous individuals over this ignoble activity. Despite these activities, there is little proof that controllers have had the option to get rid of fake prescriptions both in India and around the world [7,9].

As indicated by one gauge from BIS Research, worldwide pharma organizations lose around $200 billion to fake medications every year. Nations like India and China explicitly are truly helpless against fake items, just like a few other Southeast Asian nations who are defiling medications and placing the life and strength of their populace in impending threat. As things stand, concocting phony and terminated medications is not troublesome today – with a touch of scouring liquor, you can undoubtedly expel the ink-stepped subtleties and be good to go. As a shopper, this brings up large issues: what is medication am I purchasing, and is it valid or a sham? As things stand today, the utilization of track-and-follow inside India's pharma industry is restricted, so the open door for forging stays all the way open. We have to have controllers command the utilization of track-and-follow innovation and organizations need to firmly consider the utilization of blockchain, an innovation development frequently connected with the fluffy universe of cryptographic forms of money, for example, bitcoin, to make the pharma supply chain safe [10,15].

Blockchain is a protected decentralized framework where value-based or authentic information is recorded, put away, and kept up over a distributed system of Personal computers (PCs) called hubs. It is an unchanging, open advanced record [9]. For the pharma business, this is of gigantic intrigue, because blockchain is a virtual or advanced record that contains constantly refreshed, time-stepped, and profoundly scrambled virtual records [10]. By giving each strip, jug, or vial of medication an extraordinary number, it tends to be firmly tracked up and down the supply chain. The organizations appear to concur: 40% of well-being, executives see blockchain as among their main five needs, with spending on this innovation expected to hit $5.61 billion by 2025, as indicated by a report by BIS Research.

The reception of the blockchain innovation could spare the medicinal services industry up to $100–$150 billion every year by 2025 in information penetrate-related costs, IT costs, tasks costs, bolster work expenses and staff costs, and through a decrease in cheats and fake items, the organization reports. In India, NITI Aayog has picked pharma (and diminishing fake medications), as one of the center territories for idea reads for blockchain. "Research and meetings led … found that medications coming straightforwardly from the producer's office are reliable … the danger of passage of phony medications emerges when the items are given off between the different stages and layers of the complicated supply chain (for example wholesalers, merchants, or sub-merchants)", a NITI Aayog report called – blockchain the India Strategy, notes. At each moving point from the processing plant to the patient, medications can be taken, debased, and supplanted.

"The National Informatics Center has structured and actualized another framework named Drug Authentication and Verification Application (DAVA), in light of the GS1 norms, for medicine tracking and detectability", the report includes. "(In any case) it doesn't guarantee perceivability of every exchange to all partners … it can't track and follow the item all through the supply chain". With rising advances, for example, blockchain and Internet of Things (IoT), this can be accomplished. For customers, the execution of such a blockchain arrangement could be transformative. On the off chance that you get a case of prescriptions, you could content a novel number and get the subtleties of the medication in your grasp (maybe an image as well), with a total name and time of assembling and expiry.

The administration or controller can defend the soundness of its residents by guaranteeing the item subtleties that are entered on this circulated record and along the supply chain; its way is refreshed to guarantee total security and dispose of falsifying medicine. For a pharma organization pioneer, the enthusiasm for this sort of advancement innovation is more profound. For instance, I accept that the utilization of developments, including blockchain, can change the supply chain entirely not only limited to end user (patient), rather in reverse to keep a close track of the entire manufacturing method. Think about this situation: as organizations start to use inno-vation all the more intensely (blockchain with cloud and IoT), you can follow quality and creation measurements for each drug batch that is made, the shipping quantities, and the conduct of each medication.

10.4 LITERATURE SURVEY

In this chapter, the main focus was to study the impact and use cases of blockchain in the healthcare industry and we will lay detailed stress on blockchain drug supply chain use cases. IoT and networking, insurance, payment and banking sectors, supply chain management, online music, energy management, health, charity, real estate, retail, online data storage, private transportation, and voting are few industries that have seen blockchain use cases.

A DLT blockchain platform – MedRec – is presented in [6]. It is a platform where the records of patients are maintained and can be shared securely over the Internet. It is a bitcoin-enabled application that helps in overcoming problems such as fragmentation, system potential, and the slow access of medical records in the patient's organization. With this platform, the patients have full rights to check their medical data and take a decision on whether or not to share their data with requesting firm or party.

The block data of MedRec provide the information regarding the owner of the record and viewers that are allowed by the owner to access the records. MedRec's working mechanism makes use of the proof of work (PoW) consensus algorithm in its platform which is smart contract enabled [7]. The use of cryptographic hash in smart contracts makes the electronic health records (EHRs) of a patient more secure and increases their privacy. The correct agreement structure of MedRec is classed into the following components – patient–provider relationship (PPR) contract, sign-up contract, and a description agreement. The sign-up contract task is to identify the registrations made in the framework, the task of PPR is to gain access to the information regarding ownership, mining bounties, health record queries, etc., and descriptive agreement aims to securely maintain the medical history of a patient and how well the patient is responding to the system at present.

A major issue that is currently prevailing in the pharmaceutical industry is that these firms have records of patient data that are broken into multiple parts and are often more in number. A solution to these problems is provided in [8] in the form of a blockchain-enabled platform – Medicalchain. Only time-bounded access is provided to the health records of a patient in Medicalchain through smart contracts. The labo-ratory results and prescriptions by doctors are recorded in the form of transactions in Medicalchain. The movement of drugs all over the supply chain is recorded on the blockchain. Only time-bounded access is provided to the repository funds to settle the payments and to properly verify the treatment given to them.

The use of smart contracts makes it feasible for the patients to grant access regard-ing their medical records to the doctors so that they can provide their views on the treatment they are going through. The patients are eligible to provide access to their medical data so that they can keep track of the progress of health of a patient and even reward them with a reduction in insurance premiums or tokens. Rewards are also provided to patients for sharing their medical data with the research organizations only for some limited duration.

The patients are provided with tokens known as MedTokens by Medicalchain. These tokens can be used by patients to store information by fitness devices, to transfer payments, and to store their medical data on the blockchain. In the future, this platform will act as a motivation for various developers to make decentralized applications that would enable researchers to make use of these EHRs and provide suggestions or recommendations regarding basic hygiene and health routines.

In [9], another blockchain application – MeFy – is presented which is a subscription-based model. In MeFy, the users are allowed to sign up a yearly contract of subscription making them eligible for any sort of tests for the whole year costing them merely the price of medicines they will consume. MeFy provides users a feature of electronic consult that allows doctors from anywhere in the world to treat patients outside their location boundaries. MeMe edge device on MeFy is powered by artificial intelligence. This device will allow doctors to prescribe treatments to patients without needing the patient to visit a doctor. The doctors will utilize the patients' medical history, the treatment they have taken in past, and how effective the treatment was to auto-prescribe medicines to the affected person.

In [10] a blockchain project, MedicoHealth is presented which is designed to remove shortcomings from the pharmaceutical system. MedicoHealth allows the consumers to securely and anonymously communicate with known physicians of the world. The information about the physicians such as their area of expertise and the validity of their license is frequently updated and stored in a blockchain database. The doctors can anonymously have time-bounded access to patient's medical information. The payments made on the system are fully tokenized and are anonymous. The whole system runs on tokens.

MediBloc [11] is a platform designed on the blockchain that grants access to healthcare information for researchers, patients, and providers. This platform aims to contour drugs for researchers, patients, and providers by providing value to the data owners. MediBloc will provide a secure medical data sharing platform where the patients can rightfully own their data and the manufactures can provide information and awareness regarding their medicines and allowing the researchers to access patient's records and deduce advancement in the pharmaceutical industry.

The challenges faced by blockchain in the healthcare sector are highlighted in [12]. As per the survey, an industry that requires special focus is the healthcare supply chain industry. It is founded in an analysis by the World Health Organization that every year the global market experiences counterfeit, fake, or sub-standardized drugs of nearly about $200 billion [3]. Thus, many blockchain platforms are developed in the pharmaceutical industry to enhance the security of the drug supply chains; some of them include BlockVerify [12] and chronicled [13]. Another platform in this list of developments is farm trust [14].

In [15], the author proposed a method for secure sharing of health records between various pharmaceutical organizations. This uses a decentralized drug management system that makes use of public and private keys for security purposes. The prescription is written by the doctor after he checks the patient. The prescription provided to the patient is encrypted through the public key of the patient [17].

The medical records of patients are also encrypted and no one can access them without acquiring the patient's private key. The doctor can view the patient's health records only with the patient's permission. The use of the public and private key for encryption purposes makes the system more secure and reliable for the exchange of records.

In [16], a smart application – Healthcare Data Gateway – is proposed. It is a blockchain application that acquires a traditional database and a gateway. The working of this application is divided into the following layers – the storage, data management, and data usage layer.

The task of the storage layer is to provide secure storage of healthcare data. The security concerns are achieved by using a private blockchain, these private blockchain makes use of various techniques for the encryption of data. The work of the data management layer is to manage the data traversal access. It keeps track of all the incoming and outgoing data from the application. Lastly, the work of the data usage layer is to maintain the list of users that are accessing the patient's health records.

HealthBank is a digital start-up based on blockchain that focuses on secured and manageable sharing of medical health data between various healthcare departments [17]. This is a smart application that enables the sharing of patient's medical data securely only with the patient's authorization to do so. This application is decentralized and is highly efficient and secure.

A project was recently started by Hyperledger that was titled the "counterfeit medicine project" [18]. The main aim of this venture by Hyperledger is to detect fake drugs. Many companies are working in this direction to reduce the supply of counterfeit medicines in the global market. Some of these companies are Cisco, Accenture, Bloomberg, IBM, and Blockstream.

The drugs manufactured are provided with a time stamp in this project. This time stamp associated with each drug makes the successful tracing of the journey of the drug on blockchain [19].

A serious problem encountered by most hospitals nowadays is the unavailability of platforms that are secure and reliable for performing analysis and storage of health records. Blockchain helps in overcoming these issues by providing distributed and nontamper platform for the storage and sharing of patient's health data with various medical experts and doctors [20].

The presence of these advanced features in blockchain helps in enhancing the security of records and improving the timeliness and accuracy of diagnosis. Gem [22] is a company that focuses on incorporating the medical data of patients on the blockchain. Tierion [21] is another company that works in the same field.

In the supply chain, blockchain technology helps in safety monitoring and permanent recording of the transactions [10,15]. This majorly lessens the time consumption and thus helps in reducing man-made errors. Another prominent prevailing application of blockchain in supply chain industries is to monitor waste, cost, and labor at each step [24].

The presence of DLT helps in verifying the ethnicity and tracing the origin of the health products and thus enabling fair trade of the items. Sukhchain, BlockVerify, Provenance, and Fluent are some of the start-ups which use blockchain to improve the networks of the supply chain [4,8].

Over time, various solutions have been provided to deal with the problem of secure transfer of EHRs over cloud environments [10,15]. The work in [6] proposed an attribute-based encryption (CP-ABE) prototype. To achieve fine-grained access control for secure sharing of EHRs on the cloud, an attribute authority is employed for granting keys for data consumers. This study ignored the capability of decentralized access.

The benefits of blockchain technology in the healthcare industry are outlined by Asad Ali et al. (2019). The researcher mentioned that by introducing effective diagnosis, this technology has changed the traditional healthcare model. Researchers also note that healthcare using blockchain helps secure the exchange of information between different entities [6].

Muthanna et al. planned an associate rule to use a blockchain mechanism in IoT devices for decentralization in an exceedingly trusting manner [9]. In a project called Origin Chain, Xiwei Xu et al. executed a blockchain system. When utilizing this product, traceable information that included high accessibility was clear and tamperproof. The auxiliary plan of the framework majorly affects the nature of the system. Paula Fraga-lamas et al. suggested the use of blockchain technology in the automotive industry with a focus on cybersecurity. The exploration has given rise to an opportunity to create a new business model and may even cause a car-sharing disruption. For the same purpose, the SWOT analysis was carried out with some recommendations and future developments [15].

The author has proposed a system capable of securely sharing medical data between entities build on a hospital's private blockchain to modify the hospital's e-health system [12]. Various security properties are met by the system such as openness, tamper resistance, and decentralization, creating a safe system for the doctors to store and access medical data of patients while maintaining privacy preservation. In [17], a decentralized technique is proposed for managing EHRs. In this system, the patients can easily access their medical data across the providers. It also provides secure sharing of medical data along with managing confidentiality. Another model ProvChain is presented in [17]. The proposed framework utilizes the Merkle tree structure that helps in keeping up the security of data. The time stamp idea is utilized for the approval of data.

Provenance [19] is another blockchain arrangement that works for the detectability of items. The items can be followed all through the supply chain from producer to end user with the assistance of a unique ID doled out to every item.

The work [19] provided a system – MedShare; it is a blockchain-based platform that claims that they are adequately capable of tracking the behavior of the data and abolish access to breached permissions on data with the utilization of an access control system and smart contracts [20]. MedShare provides trailing, data provenance, and auditing on medical data that enable sharing of medical records between untrusted parties.

10.5   NEED OF BLOCKCHAIN IN PHARMA SUPPLY CHAIN

Even though viable supply chain management is the greatest threat in each industry, in the medical industry, there is extra hazard and unpredictability as an undermined supply chain in medicinal services can influence the patient's well-being [19,21].

Expanded endorsement of innovation in technology, globalization, and industry populated with different partners in different wards have offered rise to the sophisticated healthcare supply chain. The medicinal supply chain is one of the most eye-catching verticals with regards to the harmed regions of the healthcare supply chain.

One of the studies by Fraser Institute says that overall pharmaceutical deals came to USD 1.1 trillion in the year 2015. The Organisation for Economic Co-operation and Development (OECD) has seen the fake merchandise as represented 2.5% of the worldwide pharmaceutical medication businesses.

Specialists have assessed the sale of fake medications to be twice the pace of legitimate pharmaceutical exchange, which itself is a severe issue [17]. The pharma blockchain holds the possibility to upgrade security, respectability, information provenance, and the usefulness of the supply chains with its straightforward, permanent, and auditable nature.

Blockchain-centered organizations are attempting to rebuild traceability and liability in the delivery of products.

Blockchain in pharma supply chain could not just lessen duplicating and robbery issues yet in addition help oversee stock. In the worldwide healthcare domain, associations like USAID, Global Fund, or the Red Cross mean to trace back the appropriation of contributed drugs across various nations [17].

We will clarify how blockchain can be actualized in the pharma supply chain and what benefits it can bring to the biological system. Figure 10.3 explains how Drug distribution process in pharma supply chain works.

To assist individuals with understanding the capability of blockchain in the pharma supply chain, we have talked about the accompanying themes:

- Why the need for an innovative solution arises in the pharmaceutical supply chain?
- Advantages of implementing blockchain in the health supply chain.

Figure 10.4 represents the basic issues alongside traditional supply chains and how blockchain helps in solving these issues.

FIGURE 10.3 Drug distribution process in pharma supply chain.

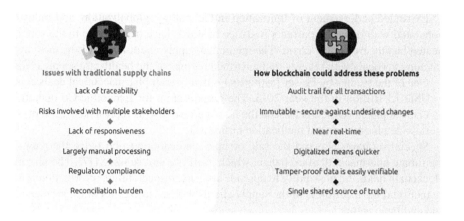

FIGURE 10.4 How blockchain can address issues hampering the supply chain.

10.5.1 MOTIVATION

As per the Supply Chain Council, the management of the supply chain comprises the management of demand and supply, sourcing crude materials and parts, warehousing and stock tracking, assembling and manufacturing, circulation over all channels, order management and order entry, and conveyance to the client. Without a doubt, the administration and coordination of the supply chain are at the center of any industry that fabricates and disseminates merchandise [17].

Simultaneously, it is understood that supply chains of various ventures are divergent as they address various requirements. For instance, in the PC business, the intensity of supply chains is tackled to offer clients item design adaptability easily; on the other side in the shopping merchandise industry, the emphasis is on item variety, accessibility, and cost. Truth be told, it very well may be handily contended that great business execution is predicated on the arrangement of a proficient supply chain; in dynamic organizations, the incorporation of supply chain management with key arranging is absolute and irreversible.

Being so crucial to the accomplishment of a business, it is just common that organizations constantly look for better approaches to arrange their supply chains to stay serious. Presently, with the guide of advancement progress made in the region of information technology, organizations are sending progressively complex solutions for additionally improving the effectiveness of supply chains.

10.6 OVERVIEW OF ELECTRONIC HEALTH RECORDS

As per the current scenario, the number of diseases and patients is continuously increasing. Therefore, in the healthcare industry, the maintenance of health records gains utmost importance. The transfer of health data across different organizations faces two major issues – the integrity issue and the privacy of data [17].

Problems associated with the storage of health data records:

Three major problems associated with the storage of health data are physical, ethical, and logistic problems.

1. **Physical problems**: Consumers consider EHRs as delicate, i.e., they do not require any physical space for storage. But this is not true, and these records must exist somewhere. Another physical issue is related to the security of health records.
2. **Ethical problems**: To overcome the issue of server space, most health record management professionals are switching to cloud computing. Some major challenges faced by cloud data storage are reliability, data portability, and integration.
3. **Logistic problems**: One of the major challenges faced in the storage of e-health records is unwanted healthcare information which means that the data can be accessed by any healthcare provider who might have not even taken any steps for the data collection.

To overcome these challenges, blockchain is being adopted. Recently, it is seen that the employment of blockchain technology in medical and healthcare services is increasing at a rapid rate. Blockchain with its secured nature has been adopted in various e-health sectors such as data access management among medical entities and secure sharing of EHRs [2].

In the case of blockchain, there is no central point of failure as the data are distributed and are stored in blocks. Blockchain technology helps in overcoming security problems in healthcare.

It has always been difficult for the pharmaceutical industries to receive support from general patients regarding sharing their clinical data with them. One major reason for patient's disapproval for sharing their medical data is a lack of security or trust. According to a survey which was conducted in 2017, nearly 32 out of 126 patients are readily approved on sharing their medical data for clinical trials so that other similar patients are benefitted strongly and timely. The results of the survey are displayed in Figure 10.5. This shows the awareness among patients to help other patients positively.

The medical records of patients help regulators in understanding valuable information regarding drugs used in clinical trials. It helps them in understanding how effective the drug is in treating a particular type of disease [6]. The results from the clinical trials have a huge impact on deciding how feasible the drug is for consumption for the patient.

10.7 ADVANTAGES OF IMPLEMENTING BLOCKCHAIN IN HEALTH SUPPLY CHAIN

There are many advantages of utilizing blockchain-enabled health product supply chains; some of them include an end-to-end tracking of health items, efficient recall management, transparency to improve responsibility, reduced misfortunes associated with counterfeiting. These are discussed in detail in the following:

- **End-to-end tracking of health items**:
    The utilization of a pharma blockchain-based arrangement will empower smoothed out discernibility of movement and partners through which medications or meds travel in the supply chain. The improved visibility encourages the enhancement of flows of products and an effective stock

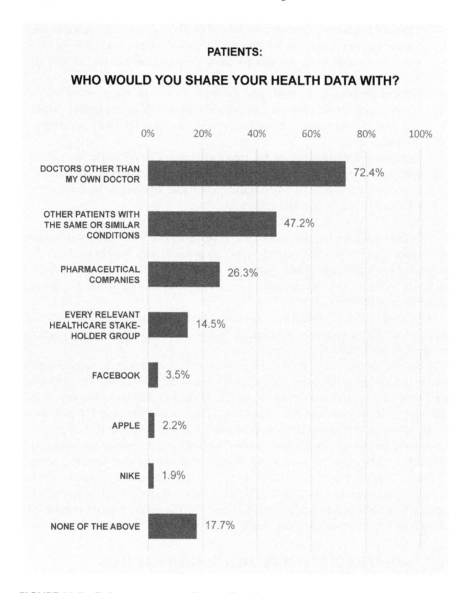

PATIENTS:

WHO WOULD YOU SHARE YOUR HEALTH DATA WITH?

FIGURE 10.5   Patient survey regarding medical data sharing.

administration framework. Figure 10.6 depicts how Blockchain could track
drugs through the supply chain and help wholesalers access credit.

- **Reduced misfortunes associated with counterfeiting**:

  A blockchain-enabled application is capable of keeping track of the
  movement of healthcare products throughout the supply chain beginning
  from manufacturer to the end user. Thus, it would get conceivable to inspect
  weak focuses in the supply chain and lessen the odds of frauds and the
  expenses related to it.

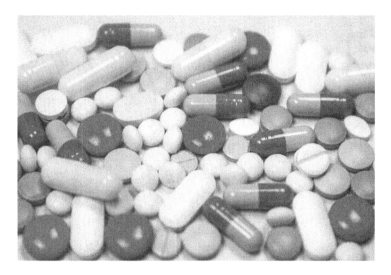

FIGURE 10.6 Blockchain could track drugs through the supply chain and help wholesalers access credit.

- **Improved responsibleness through transparency**:
  The traversal details of medicine in the drug supply chain are stored on the blockchain and can be traced at any time easily. If any corruption is signified in the delivery of drugs, then blockchain data can be used to trace the source of the last owner of the medicine or the one responsible for duplicate meds.
- **Effective management of recall batches of medicines**:
  The use of blockchain has enhanced the security and reliability of the supply chain. Now, the location of the drugs can easily be tracked. Another efficient use case is that the data stored on the blockchain can help firms to get timely reminders regarding the stock of a particular medicine and thus a bunch of medicines can be reordered before finishing. Figure 10.7 depicts the key concepts of blockchain-enabled drug supply chain.

10.8 TAMPER VERIFICATION SUPPLY CHAINS (AND OTHER GAME-EVOLVING OPENINGS)

Blockchain is quickly rising as one of the most significant advancements to deal with the majority of information engaged with drug turn of events and supply – from clinical preliminaries to buyer advertising to logistics. Blockchain was created to track and record bitcoin exchanges and it makes use of a model comprising distributed ledger.

Since data are imitated over each hub in the system, the innovation responds like the carefully designed seals on current drugs, as it makes it practically difficult to misrepresent data put away on the blockchain.

One of the quickly evident employments of blockchain in the supply chain is to enlist the exchange of products between two gatherings, recognized as two locations

Control	**Inferred Trust**	**Flexibility**	**End to End visibility**
Arrangements, checking Abilities, control Instruments are of prime Significance	The partners ought not need to believe one another yet should believe the security coordinated into the framework to gather and give genuine and precise information	The capacity to adjust and react to issues quickly with no basic effect on operational expenses are important to remain competitive in the market	Observing supply chain network occasions and procedures evaluate lament focuses and recognize issues and issues, even proactively. This idea is vital to expand productivity along the supply.

FIGURE 10.7   Key concepts of blockchain-enabled drug supply chain.

in the blockchain. The exchange, including all supporting data, is then signed in the blockchain and is made accessible to all gatherings engaged with the exchange, empowering them to follow its provenance all through the whole supply chain.

Blockchain's dispersed record implies that it is not possible for anyone to control this information, enabling controllers to decide with sureness who is answerable for defilement or another consistency penetrates and where these happened.

Blockchain cannot simply build recognizability and battle forging at the supply level; it likewise can reform the whole drug supply chain from manufacturers to patients, warehousing, shipment through to conveyance, and even at the drug store, specialist, clinical preliminary level, and patients.

In any case, the advantages are not absolutely about monitoring drugs, significant as that seems to be. There are a large group of other potential applications, from making clinical information carefully designed to making administration and consistency increasingly straightforward, to utilizing blockchain to assist wholesalers with getting to credit quicker through progressively productive credit checking and onboarding.

10.9   PROPOSED SYSTEM

10.9.1   PROBLEM STATEMENT

One of the major issues faced by pharmaceutical industries nowadays is falsified medicines or fake drugs.

The following are considered as per the World Health Organization estimates:

- An evaluated 1 out of 10 meds in underdeveloped and developing nations is either counterfeit or inadequate.
- At least 100,000 to 1 million individuals lose their life each year because of fake medications.
- It is evaluated that at any rate 72,000 kids pass on of pneumonia and 69,000 individuals kick the bucket of intestinal sickness every year because of adulterated or unacceptable medications.
- In nations like the USA, the administration has taken vital measures to secure drugs, yet at an expense of 27 billion dollars per year.
- It is nearly impossible to detect or trace the origin of these counterfeit meds in the supply chain as they have to traverse through a very complex network going through many entities in the middle. This lacking grants doorways for falsified or fake medicines to enter the worldwide market. This makes the security of the supply chain system of utmost importance to today's systems.

10.9.2 Objective of Research

The objectives of the dissertation may be summarized as follows:

- A survey of the historical research work on blockchain applications in the healthcare and medicinal field will be performed.
- Summarize and categorize existing benefits/challenges on incorporating blockchain in the healthcare domain.
- Exploring new research activities for forming a roadmap for the future.
- Implementation of a blockchain-based framework for storing drug transactions taking place throughout the supply chain from manufacturer to consumer.

Why do we need blockchain in the pharmaceutical value chain? The answer to this question can simply be given through various features of blockchain which make it secure and reliable for the storage of data. Two such features are that the blocks in the blockchain are immutable and time-stamped, i.e., the data stored on the blockchain remains the same [15]. These features make blockchain suitable for implementation in the pharma supply chain [4].

The participating companies can choose between a private or public blockchain based on their requirements. Only limited access will be provided as per the regulations of the smart contract for sharing of data agreed upon by the parties. To keep track of the participating entities during the drug distribution process, the organizations use distributed ledgers among them.

The use of blockchain makes monitoring drugs throughout the drug supply chain an easily achievable task, thus helping in reducing the number of falsified medicines from the market [26]. The viable process of drug movement can entirely be recorded on blockchain including the beginners or drug makers to the last party being the affected person [4].

Some important measurable that will be achieved by blockchain-enabled supply chains in healthcare is that the clinical laboratories will be able to take timely actions by identifying the source of ambiguity and the companies will be able to trace the medicines in the supply value chain [27]. This will help in creating a secure circuit stopping the entrance of fake medications in the chain of supply.

In Figure 10.8, the system architecture represents stakeholders involved in the drug supply chain. Each member of the system has their specific roles. The roles of these stakeholders can be distinguished as follows:

1. **Manufacturer**: The role of the manufacturer involves taking orders from the pharmacy and manufacture good quality drugs. The manufacturer can monitor and make a record of the units of medicines that are manufactured and available for selling.
2. **Distributor**: Buying medicines from the manufacturer and shipping them for sale to retailers comprises the role of distributor. The distributor can transfer ownership of medicines to the retailers.
3. **Retailers**: The role of retailers involves buying medicines from the distributors and sell them to the consumers or end users [25]. The retailer's account can transfer ownership of medicines to the consumers.
4. **Consumers**: Buying medicines from the retailers and verifying the origin of medicines to evaluate the authenticity of drugs are the roles of consumers in the drug supply chain.

10.9.3 Modeling Drug Distribution Process

Stakeholders that are associated with the drug supply chain are as follows:

- Manufacturer
- Distributor
- Retailer
- Consumer.

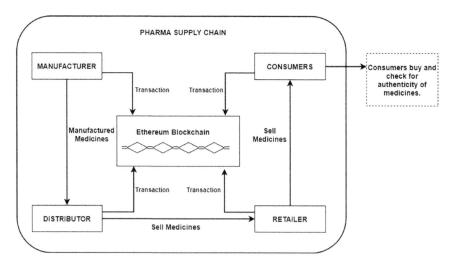

FIGURE 10.8 System architecture.

MANUFACTURER

- The manufacturer manufactures drugs and assigns it with a unique code.
- The information is stored on the blockchain

WHOLESALER

- The Wholesaler verifies the origin of the product/drug.
- The transaction is added to the blockchain.

PHARMACIST

- The Pharmacist also verifies the origin of the product.
- The transaction between wholesaler and pharmacist is stored to the blockchain.

PATIENT

- The Patient verifies the origin of the product
- The transaction between pharmacist and patient is added to the blockchain.

FIGURE 10.9   Medical supply chain storage flow for drug safety.

1. The producer fabricates the meds and stores the exchange on the blockchain (Figure 10.9).

   The first step in this process is the creation of medicines by the manufacturer. The producer makes the medications and adds this transaction to the blockchain. The transaction comprises basic details regarding the manufacturing such as manufacturer ID, location, expiry date, assembling details, and date of manufacturing [26].

   The other stakeholders can easily verify the authenticity of the origin of medicines by tracing the details stored by the producer on the blockchain regarding the manufacturing details of the drugs.

   A hash ID is generated when the information is stored on the blockchain; this hash ID can be used for tracing the drugs for manufacturer detail verification [27]. Cold chain transporting is another implication of blockchain in the health value chain. Environment-sensible medicines can be transported using IoT vehicles capable of maintaining records through temperature sensors installed within [4,5].

The data of these sensor-enabled vehicles can be stored on the blockchain. This will help laboratories in identifying whether the temperature-sensible medicines are delivered accurately or not. If a faulty medicine is delivered then blockchain data can help us in identifying the person responsible for it.

The location details can also be shared by these IoT-enabled vehicles to the blockchain. The aim of sharing their location is to help the government and other stakeholders in verifying the interval of time at which these medicines reach the person or organization ordering them [8,5].

2. The distributors order the medicines from manufacturers and send them to retailers.

   The distributors on receiving the medicines from manufacturers can verify and ensure their authenticity. This can be done by using the hash ID for tracing details of the manufacturer and basic details stored on the blockchain.

   The distributors can verify the details of medicines such as the date of manufacturing, location of manufacturing, and expiry date.

   On verifying the manufacturer details, the distributor approves the transaction and again stores it on the blockchain.

   After this, the distributors send or sell the medicines forward to the retailers who are involved in the supply chain.

3. The retailers receive the medicines from distributors and check their source of origin.

   The retailers on receiving the medicines can verify the origin details through the hash ID that is uniquely generated for every transaction.

   Supposedly if a distributor tries to sell falsified medicines to the retailers using a fake hash ID. Then, the retailer can verify the hash ID and declare the transaction as invalid due to the presence of counterfeits.

   Moreover, over a blockchain platform for the supply chain, only valid participants are allowed to access block data.

   Similarly, the drug authorities can be made aware of the fake drug manufacturers when they try to enter a secure blockchain supply chain.

   The drug retailers on verifying the origin of medicines approve them and store the transaction regarding the distributor and retailer on the blockchain. Now, the patients are allowed to buy drugs from the retailers.

4. The consumers or end users on receiving meds from retailers verify its source.

   Consumers can guarantee themselves with surety regarding the authentic nature of medicines that they receive from retailers. They can verify whether the medicine they are going to consume is legitimate or not. This can be done by tracing the source of received medicine through the hash ID printed on the tablet's bottle. They can even verify whether the medicine meets quality standards or not [8].

   The hash ID would bring data from the blockchain, and patients can get to the fundamental medicine details.

The patients are allowed to share their views regarding the effectiveness of ordered medicine on the blockchain so that company can get ideas regarding if they need improvements or not.

The patient can even provide ratings to medicines which in turn will help in deciding the power of medicine in terms of its positive effects.

In this way, the entire process of drug manufacturing can be monitored using Ethereum, which is a blockchain-based distributed computing platform that features smart contract functionality. Every time the drug traverses from one stakeholder to another; the transactions are stored securely on the blockchain. This helps in monitoring the drug movements from manufacturer to end users and overcome the problem of counterfeits from the pharmaceutical supply chain.

10.9.4 Limitations of Using Blockchain in Drug Supply Chain

* Limited processing power and scalability:
  Most of the versions of blockchain available nowadays are less scalable and very slow. Hyperledger and Ethereum both require a great amount of time for a new block to get added to the blockchain. Thus, emphasis should be given to generating blockchain versions that are highly scalable and time-effective.
* Complexity in terms of usability:
  Blockchain is still very new and understood by only a few people. Thus, it becomes difficult for everyone to work on its complex structure.
* There is a need for the development of a new paradigm that requires a holistic approach and thinking.
* The return on investment is unknown.
* Blockchain does not have any validation and standardization.
* The project owner needs to bear the costs for storing records on blockchain for a longer period.
* Deletion of data is not allowed. You can only update the data or verify it.
* The management of private keys is very complex for now which results in the loss of digital ID over the supply chain.

10.10 CONCLUSION AND FUTURE SCOPE

Blockchain enhances the security of the pharmaceutical supply chain by storing drug movements over a tamperproof platform. The proposed system provides a blockchain solution for monitoring the pharma drug supply chain and reducing counterfeits to great extent. The proposed framework guarantees drug security as well as manufacturer's authenticity in the pharma supply chain. Some future directives can be toward proper traceability of drugs and patient's clinical history management. Adding the clinical history of patients on blockchain will allow doctors to grant proper prescriptions to patients from anywhere in the world. Blockchain can overcome various issues prevailing in the supply chain of the pharmaceutical industry.

REFERENCES

[1] Tyndall, G., Gopal, C., Partsch, W., and Kamauff, J. (1998) "Supercharging supply chains", In *New Ways to Increase Value Through Global Operational Excellence*.

[2] Arsene, C. (2019) Hyperledger Project Explores Fighting Counterfeit Drugs with Blockchain. Available online: https://healthcareweekly.com/blockchain-in-healthcare-guide/, accessed on March 20, 2019.

[3] el Maouchi, M., Ersoy, O., and Erkin, Z. (2018) TRADE: a transparent, decentralized traceability system for the supply chain. In *Proceedings of 1st ERCIM Blockchain Workshop 2018. European Society for Socially Embedded Technologies (EUSSET)*.

[4] Saini, K. (2018) "A future's dominant technology blockchain: digital transformation", In *2018 International Conference on Computing, Power and Communication Technologies (GUCON)*, IEEE.

[5] Release, I. N. (2017) Maersk and IBM Unveil First Industry-wide Cross-border Supply Chain Solution on Blockchain [Online]. Available from: https://www-03.ibm.com/press/us/en/pressrelease/51712.wss#feeds.

[6] Allison, B. I. (2016) Shipping Giant Maersk Tests Blockchain-Powered Bill of Lading [Online]. Available from: http://www.ibtimes.co.uk/shipping-giant-maersk-tests-blockchain-powered-bills-lading-1585929.

[7] Kumari, K., and Saini, K. (2019) "CFDD (Counterfeit Drug Detection) using Blockchain in the Pharmaceutical Industry", *Int J Eng Res* 08(12), 591–594.

[8] Pant, R. R., Prakash, G., and Farooquie, J. A. (2015) "A framework for traceability and transparency in the dairy supply chain networks", *Procedia-Soc Behav Sci* 189, 385–394.

[9] Dutta, S., and Saini, K. (2020) "Blockchain and social media", In *Blockchain Technology and Applications*, pp. 101–114, Auerbach Publications, Boca Raton, Fl.

[10] Saini, K. (2021) "Next generation logistics: a novel approach of blockchain technology", In *Essential Enterprise Blockchain Concepts and Applications*, pp. 143–152, Auerbach Publicatio, Boca Raton, Fl.

[11] Pilkington, M. (2016) "11 Blockchain technology: principles and applications", In *Research Handbook on Digital Transformations*, vol. 225, London, UK: Edward Elgar Publishing.

[12] MedicoHealth. (2018) MedicoHealth, MedicoHealth Whitepaper. Available online: https://medicohealth.io/supporters/documents/wp_beta.pdf, accessed on March 15, 2019.

[13] Ekblaw, A., Azaria, A., Halamka, J. D., and Lippman, A. (2016) "A case study for blockchain in healthcare: "MedRec" prototype for electronic health records and medical research data", In *Proceedings of the IEEE Open & Big Data Conference*, vol. 13, p. 13, Vienna, Austria.

[14] Chang, Y., Iakovou, E., and Shi, W. (2019) "Blockchain in global supply chains and cross border trade: a critical synthesis of the state-of-the-art, challenges, and opportunities", ArXiv preprint arXiv:1901.02715.

[15] Kavita, K., and Saini, K. (2020) "Data handling & drug traceability: blockchain meets healthcare to combat counterfeit drugs", *Int J Sci Technol Res* 9(3), 728–731.

[16] Mettler, M. (2016) "Blockchain technology in healthcare: the revolution starts here", In *Proceedings of the 2016 IEEE 18th International Conference on e-Health Networking, Applications and Services (Healthcom)*, pp. 1–3, Munich, Germany.

[17] Pethuru, R., Saini, K., and Surianarayanan, C., (Eds.) (2020) *Blockchain Technology and Applications*, CRC Press, Boca Raton, Fl.

[18] MediBloc. MediBloc, MediBloc Technical Whitepaper. (2018) Available online: https://github.com/medibloc/whitepaper/blob/master/old_whitepaper/medibloc_whitepaper_en.pdf, accessed on March 15, 2019.

[19] Kavita, S. (2019) "Evolution of blockchain technology in business applications", *J Emerging Technol Innov Res (JETIR)* 6(9), 240–244.

[20] Deimel, M., Frentrup, M., and Theuvsen, L. (2008) "Transparency in food supply chains: empirical results from German pig and dairy production", *J China Network Sci* 8(1), 21–32.

[21] Kshetri N. (2018) "1 Blockchain's roles in meeting key supply chain management objectives", *Int J Inf Manage* 39, 80–89.

[22] Azaria, A., Ekblaw, A., Vieira, T., and Lippman, A. (2016) MedRec: using blockchain for medical data access and permission management. In *Proceedings of the 2016 2nd International Conference on Open and Big Data (OBD)*, pp. 25–30, Vienna, Austria.

[23] Medicalchain. Medicalchain, Medicalchain Whitepaper 2.1. Tech. Rep. Medicalchain. (2018) Available online: https://medicalchain.com/Medicalchain-Whitepaper-EN.pdf, accessed on March 14, 2019.

[24] Poirier, C. C. (1999) *Advanced Supply Chain Management: How to Build a Sustained Competition Advantage*, Berrett-Koehler Publishers.

[25] Medicalchain. MeFy, MeFy Whitepaper. (2018) Available online: https://icosbull.com/whitepapers/3576/MeFy_whitepaper.pdf, accessed on March 15, 2019.

[26] Saugata Dutta, Kavita Saini, "Securing Data: A Study on Different Transform Domain Techniques", WSEAS Transactions on Systems And Control, Volume 16, 2021, E-ISSN: 2224-2856, DOI: 10.37394/23203.2021.16.8

[27] Saugata Dutta, Kavita Saini, "Securing Data: A Study on Different Transform Domain Techniques", WSEAS Transactions on Systems And Control, Volume 16, 2021, E-ISSN: 2224-2856, DOI: 10.37394/23203.2021.16.8

Index

Printed in the United States
by Baker & Taylor Publisher Services